A Text Book of Clinical Toxicology

T. S. Mohamed Saleem M. Pharm., (Ph. D)

Assistant Professor & Head

Department of Pharmacology

Annamacharya College of Pharmacy

Andhra Pradesh, India

LULU PUBLISHING HOUSE

Published by

Lulu Publishing House

e-mail: Visit our website: www.lulu.com

A Text Book of Clinical Toxicology

© 2013, T.S. Mohamed Saleem

All rights reserved. No part of this publication should be reproduced, stored in a retrieval system, or transmitted in any form or by any means: electronic, mechanical, photocopying, recording, or otherwise, without the prior written permission of the author and the publisher.

This book has been published in good faith that the material provided by author is original. Every effort is made to ensure accuracy of material, but the publisher, printer and author will not be held responsible for any inadvertent error(s).

First Edition: 2013

ISBN 978-1-304-52077-7

Typeset at Lulu typesetting unit

Printed at Lulu Publishing House

I dedicate this book to
my parents, wife and daughter.

Preface

A text book of clinical toxicology presents closely and resourcefully the scientific root of poisonous effect of substance and drugs. The book covers all aspects of toxicological effect of drugs and related substance. The book represents an updated version of clinical toxicology in easy understanding language. The book was written for the students studying health related course who required more information regarding general management of poisoning and drug related adverse effects. The topics covered in the book have been carefully selected based on the curriculum design for undergraduate students especially Pharm.D. The book covers all necessary components to communicate the indispensable required skills to students in the subject of clinical toxicology as per medical and pharmacy regulation. The book covers totally 19 chapters and all individual chapters carefully discussed about the pre-hospital and hospital management of poisons and drugs. First two chapters discussed about general procedure regarding management of poisons and remaining chapters covers individual poisons and its management. I am thankful to all my students for their insight on clinical toxicology. That in a way encouraged me to write this book. I hope that students and teachers would benefit from this book. Suggestions for improvements from teachers and students are most welcome.

T.S. Mohamed Saleem

Assistant professor & Head

Department of Pharmacology

ANCP, India

Content

Chapter	Title	Page No
Chapter-1	GENERAL MANAGEMNT OF POISONING	01
Chapter-2	ENHANCED ELIMINATION	14
Chapter-3	TOXICOKINETICS	27
Chapter-4	MANAGEMENT OF PARACETAMOL POISONING	38
Chapter-5	ASPIRIN TOXICITY	44
Chapter-6	NONSTEROIDAL ANTI-INFLAMMATORY DRUGS	49
Chapter-7	TRICYCLIC ANTIDEPRESSANTS	52
Chapter-8	ACUTE POISONING OF OPIATES	55
Chapter-9	BARBITURATE AND BENZODIAZEPINE POISON	59
Chapter-10	ALCOHOL TOXICITY	62
Chapter-11	RADIATION POISONING	67
Chapter-12	ACUTE POISONING OF HYDROCARBONS	70
Chapter-13	CAUSTICS POISON	76
Chapter-14	MUSHROOM POISONING	79
Chapter-15	MYCOTOXIN	84
Chapter-16	FOOD POISONING	89
Chapter-17	SNAKE POISON	101
Chapter-18	HEAVY METAL TOXICITY	111
Chapter-19	ORGANOPHOSPHOROUS POISONING	118

CHAPTER-1

GENERAL MANAGEMNT OF POISONING

Definition

Poisoning occurs when any substance interferes with normal body functions after it is swallowed, inhaled, injected or absorbed. Drug overdoses occur when pharmacologic preparations are taken in excess and lead to toxicity.

Pre hospital management

Suspect poisoning in the following condition

- Unexplained loss of consciousness or altered sensorium
- Vomiting, difficulty breathing, sleepiness, confusion or other unexpected signs
- Burns or redness around the mouth and lips, which can result from drinking certain poisons
- Breath that smells like chemicals, such as kerosene, insecticides or any unusual smell
- Burns, stains and odors on the person, or clothing or on the furniture, floor or other objects in the surrounding area
- Empty medication bottles or scattered pills

Call the local medical emergency service if the person is

- Drowsy or unconscious
- Having difficulty breathing or has stopped breathing
- Having seizures

What to do while waiting for help

- If the person is conscious and you are sure that he has not taken kerosene or petroleum products, corrosives or heavy metals, give charcoal slurry (homemade charcoal – *burnt toast mixed with tea*).

- If the patient has taken corrosives, egg white can be safely administered.
- If the patient is unresponsive, open the airway and look for signs of breathing. If no breathing is present, deliver artificial breathing only in the presence of barrier devices or ambu-bag and check for signs of circulation (carotid pulse if you are a doctor/paramedic and cough/limb movements for non-medical personnel).
- Start CPR if there are no signs of circulation. If the victim is breathing, put the patient in recovery position (lateral position) and reassess every 2 minutes.

If the person starts having convulsions, give *convulsion first aid*:

- Protect the person from injury. Try to prevent a fall. Lay the person on the ground in a safe area. Clear the area of furniture or other sharp objects. Cushion the person's head. Loosen tight clothing, especially around the person's neck. Turn the person on his or her side. If vomiting occurs, this helps make sure that the vomit is not inhaled into the lungs. Stay with the person until recovery or until you have professional medical help.

- DO NOT restrain the person. DO NOT place anything between the person's teeth during a seizure (including your fingers). DO NOT move the person unless he or she is in danger or near something hazardous. DO NOT try to make the person stop convulsing. He or she has no control over the seizure and is not aware of what is happening at the time. DO NOT give the person anything by mouth until the convulsions have stopped and the person is fully awake and alert

SUPPORTIVE CARE IN CLINICAL TOXICOLOGY

Initial resuscitation should be based on the assessment of the patient and not the particular toxin involved and standard advanced life support (ALS) guidelines should be followed. Specific instances where treatment may differ are indicated below.

"The majority of patients taking overdoses or with drug toxicity are young and healthy, so cardiac and respiratory support should be continued for much longer periods of time in patients with a toxicity-related cardiorespiratory arrest".

If there is any doubt, cardiac compression and ventilatory support should be continued until the situation has been discussed with a clinical toxicologist. There has been survival with normal neurological function in patients receiving cardiopulmonary resuscitation (CPR) for hours.

Airway

There are no specific differences from Resuscitation except for caustic and corrosive ingestions. CNS depression is a common effect of drugs, so regular and careful assessment of airway protection and patency is important.

Breathing

Toxicology patients rarely have hypoxia unless they develop aspiration pneumonitis. The commonest problem is hypoventilation secondary to respiratory depression.

Circulation

Inotropic support

The use of intravenous fluid therapy and inotropic support should be based on patient haemodynamics and the specific toxins ingested. Although specific inotropes or or other drugs are suggested in toxicology patients, the initial management of cardiogenic shock should be the same as for any other cause unless there are specific contraindications to particular inotropes. The initial inotrope of choice is adrenaline unless its vasopressor actions are contraindicated, such as in beta blocker overdose.

Administration of an inotrope should only be undertaken in consultation with a toxicologist or cardiologist. Other inotropes are used in toxicology, but should usually be used in consultation with a clinical toxicologist.

Prolonged cardiopulmonary resuscitation is essential because unlike in arrests due to cardiovascular disease, the majority of patients are healthy prior to the overdose, and survival with normal neurological function after long periods (hours) of cardiopulmonary resuscitation is well documented.

Adult doses of other inotropes used in toxicology.

Milrinone (phosphodiesterase inhibitor)

Milrinone 50 micrograms/kg IV, slowly over 10 minutes, followed by 0.375 to 0.75 micrograms/kg/minute IV, adjusting according to clinical and haemodynamic response, up to a maximum of 1.13 mg/kg daily.

Insulin euglycaemia

1 short-acting insulin 1 unit/kg IV bolus, followed by 1 unit/kg/hour. The dose can be increased to 2 units/kg/hour or further but this should be discussed with a clinical toxicologist PLUS glucose 10% or 50% IV infusion OR 2 glucagon 5 to 10 mg (= 5 to 10 units) IV bolus, then continue at 5 to 10 mg/hour.

Dobutamine

dobutamine 2.5 to 10 micrograms/kg/minute IV.

Drug-induced arrhythmias

QT prolongation and torsades de pointes: QT prolongation should be monitored and any other precipitating factors should be determined and treated if possible. Electrolytes, including magnesium and calcium, should be checked and deficiencies corrected. Patients with hypomagnesaemia should have magnesium replacement. [In toxicology patients, the correction of magnesium deficiency, if required, needs to be done rapidly while the patient has a prolonged QT interval and is at risk of torsades de pointes]

In adults, use: **magnesium sulfate** 50% 5 to 10 mL (= 2.5 to 5 g or 10 to 20 mmol) IV over 30 to 60 minutes.

In children, use: **magnesium sulfate** 50% 0.1 mL/kg (= 50 mg/kg or 0.2 mmol/kg) IV over 20 minutes [Magnesium can be given intramuscularly in infants or children if urgent IV access not possible], followed by 0.06 mL/kg/hour (= 30 mg/kg/hour or 0.12 mmol/kg/hour) IV infusion.

Patients with hypocalcaemia should have calcium replaced. Use: In adults, use: **calcium gluconate** 10% 10 to 20 mL (= 1 to 2 g or 2.2 to 4.4 mmol) IV, over 10 to 30 minutes.

In children, use: **calcium gluconate** 10% 2 to 5 mL/kg/day (= 200 to 500 mg/kg/day) IV infusion. Patients with hypokalaemia should have potassium replaced. If the patient is able to take and absorb oral potassium, use: potassium chloride 14 to 16 mmol orally, 3 times daily (child: 1 mmol/kg/day in 2 to 4 doses) [Effervescent immediate-release tablets of potassium contain 14 mmol potassium per tablet, and slow-release tablets contain 8 mmol potassium. The slow-release formulations of potassium are almost completely absorbed within one hour].

If the serum potassium is less than 3 mmol/L or the patient is unable to take or absorb oral potassium, use: potassium chloride 10 to 20 mmol (= 0.75 to 1.5 g) IV, over 1 to 2 hours (child: 0.6 mmol/kg IV over 3 hours) preferably as a pre-mixed solution of the appropriate intravenous fluid. [If pre-mixed IV solution is unavailable, potassium chloride concentrate injection must be added to a large volume of parenteral fluid and mixed thoroughly before infusion. The usual maximum concentration is 40 mmol/L].

Isoprenaline or transvenous pacing should be considered in patients with a prolonged QT interval and bradycardia. In adults, use:

Isoprenaline 20 micrograms IV, repeat according to clinical response, and commence an infusion at 1 to 4 micrograms/minute, but the rate may need to be rapidly increased t o give double, quadruple or higher doses as required to overcome the beta blockade.

Torsades de pointes may resolve spontaneously within a minute but if not, first-line treatment is a 200 J DC shock or equivalent. If there is no response to an initial DC shock this can be repeated with increasing voltage shocks. Magnesium should also be given (except in torsades de pointes resulting from beta blocker overdose, see Toxicology: beta blockers).

In adults, use: **magnesium sulfate** 50% 2 to 4 mL (= 1 to 2 g or 4 to 8 mmol) IV as a slow injectionover 2 to 5 minutes.

In children, use: **magnesium sulfate** 50% 0.05 to 0.1 mL/kg to a maximum of 4 mL (= 0.025 to 0.05 g/kg or 0.1 to 0.2 mmol/kg to a maximum of 8 mmol/dose) IV as a slow injection over 10 to 15 minutes.

DECONTAMINATION

The most important first aid or therapeutic measure to minimize morbidity and mortality from ingest

Gastric aspiration and lavage

The commonest method used in our hospitals for this purpose is gastric aspiration and lavage (stomach wash out). Lavage is a very useful procedure in the first hour of ingestion. However, certain drugs like salicylates have been recovered many hours after ingestion.

Clinical studies have not confirmed the benefit of the gastric lavage alone, even when it was performed less than 60 minutes after ingestion of a poison. However, almost all these studies involved drug overdoses. Considering the high incidence of pesticide poisoning in Sri Lanka, gastric lavage is recommended up to 4 hours after poisoning.

If the patient is conscious, the procedure should be explained to him and his consent must be obtained. A struggling or uncooperative patient should not be subjected to the procedure, since this increases the risk of complications. Gastric lavage is not recommended outside a health care facility. It should not be employed routinely in the management and should not be considered unless a patient has ingested a potentially life-threatening amount of a poison.

The patient should be placed in the left lateral head down position. A large bore (36-40 French or 30 English gauge) tube (external diameter 12-13.3 mm), should be used in adults and 24-28 French gauge tube (external diameter 7.8-9.3 mm) in children. The lavage tube should have a rounded end. It should be sufficiently firm to be passed into the stomach, and yet flexible enough to prevent mucosal damage.

The end of the oro-gastric tube can be lubricated with water-soluble jelly. It is introduced through the mouth. The length of the tube to be inserted is measured and marked before insertion. In an adult, at about 40 cm the gastro-oesophageal junction is reached. Once passed, the position of the tube should be checked either by air insufflation while listening over the stomach, and/or aspiration with pH testing of the aspirate. The first sample aspirated should be labelled and preserved for toxicological analysis.

A volume of 10 ml/kg body weight for each cycle of lavage can be used for children, and about 200 to 400 ml for adults. Lavage should be continued until the fluid coming out is clear for several cycles. 0.9% sodium chloride is considered the best for lavage in children. For adults, use pure water or 0.9% sodium chloride. Epigastric massage to increase the return of particulate poisons is recommended by some. The volume of lavage fluid returned should approximate to the amount of fluid given. In cold climates, the fluid should be warm (380C) to prevent hypothermia, especially in children.

Lavage is contraindicated after ingestion of corrosives (concentrated acids, alkalis etc.), and petroleum products, such as kerosene. Under certain circumstances, for example, when a pesticide is dissolved in kerosene or an excessive amount is ingested for suicidal purposes, lavage may be performed for kerosene poisoning, if the patient can cough effectively, or after inserting a cuffed endotracheal tube. Lavage should not be performed in patients whose airway protective mechanisms are impaired or expected to be affected due to impaired consciousness, coma or convulsions. These patients can be lavaged following intubation with a cuffed endotracheal tube.

Gastric lavage is also contraindicated in patients who are at risk of haemorrhage or gastro-intestinal perforation due to pathological conditions, who had recent surgery or other medical conditions that could be further compromised by this procedure.

Aspiration pneumonia, laryngospasm, hypoxia, hypercapnia, mechanical injury to throat, oesophagus and stomach (including perforation), and fluid and electrolyte imbalance are some of the complications of gastric lavage. Ion is the removal of the poison from the gastro-intestinal tract.

Emesis

Emesis is another method of removing an ingested poison. It does not necessarily empty the stomach. Spontaneous emesis follows the ingestion of many irritant poisons. There is little proof that it effectively empties the stomach.

Emesis can be induced by giving syrup of ipecacuanha (Paediatric Ipecacuanha Emetic Mixture BP or Ipecac syrup USP).

Dose -

6-12 months: 5-10 ml preceded or followed by 120-240 ml of water.

1 to 12 years: 15 ml preceded or followed by 120-240 ml of water.

Over 12 years: 15-30 ml followed immediately by 240 ml of water.

In children up to six months, ipecac syrup should be administered under the supervision of a doctor.

Ipecac should be given only in an alert conscious patient, who has ingested a potentially toxic amount of a poison.

There is insufficient data to support or exclude ipecac administration soon after ingestion of a poison.

Emesis is most effective if initiated within 30 minutes of ingestion. The long time to initiate emesis is a major disadvantage of syrup of ipecac. Therefore, in severe poisoning lavage is considered the better procedure.

The contraindications for emesis are the same as for lavage. If coma and convulsions are present or anticipated, emesis should not be induced. Relative contraindications for emesis are late-stage pregnancy, age less than six months, severe cardiac disease and debility, or medical conditions that may be further compromised by the induction of emesis.

Complications of syrup of ipecac are diarrhoea, lethargy, and drowsiness and prolonged vomiting.

Salt solutions, apomorphine, copper sulphate etc. should not be used to induce emesis because of their high risk-benefit ratio.

Adsorption

Adsorption of poisons to certain substances such as activated charcoal is another method used to reduce absorption of a poison from the gastro-intestinal tract. This is widely used for drugs. It can be effectively used for pesticides and most other substances where the toxic dose is relatively small. The main substances not effectively adsorbed by activated charcoal are iron, lithium, alcohols and glycol.

The effectiveness of activated charcoal decreases with time and the greatest benefit is within one hour of ingestion.

Dose -

Up to one year: 1 g/kg. 1 - 12 years: 25 to 50 g.

Over 12 years: 50 to 100 g.

Activated charcoal suspended in 200 ml of water is administered orally or instilled via a nasogastric tube, after emesis or gastric lavage.

For paraquat poisoning, Fuller's earth is the adsorbent of choice.

Activated charcoal may also be used. Activated charcoal is contraindicated if the patient's consciousness is impaired, if he is at risk of haemorrhage or gastrointestinal perforation due to recent surgery or some other pathology. Activated charcoal is also contraindicated if its use increases the risk and severity of aspiration.

Multiple-dose activated charcoal (MDAC) therapy (repeated administration of oral activated charcoal) enhances elimination of poisons by interrupting the entero-enteric and in some cases, the entero-hepatic circulation.

MDAC can increase the elimination of carbamazepine, dapsone, dextropropoxyphene, digoxin, disopyramide, nadolol, paracetamol, phenytoin, piroxicam, quinine, sotalol and theophylline. MDAC may increase the elimination of salicylate. After an initial dose of 50 to 100 g, it is recommended that activated charcoal should be administered at a rate not less than 12.5 g per hour until recovery and/or the plasma drug concentration is within the normal therapeutic range.

Cathartics

Saline or osmotic cathartics have been used to further decontaminate the gut. The administration of cathartics alone has no role in the management of poisoned patients.

Whole bowel irrigation

Whole bowel irrigation (WBI) with an isotonic solution may be useful to empty the bowel for iron, lithium, ingested button batteries, ingested illicit drug packets and overdose of sustained release or enteric coated drugs. In WBI the intestines are cleaned by the oral administration of large amounts of polyethylene glycol electrolyte solution. This is not available in Sri Lanka. Before, during and after decontamination procedures it is essential to check that the patient is breathing properly. The airway should be clear. Secretions should be sucked out. If cyanosed, oxygen must be given. If the respiration is impaired, intubation and assisted ventilation should be considered.

ELIMINATION

Alkaline and acid diuresis, peritoneal dialysis, haemodialysis and haemoperfusion are the methods used to eliminate absorbed poisons. The common practice of giving frusemide IV for all cases of poisoning should be discontinued.

Alkaline diuresis

Alkaline diuresis increases the elimination of salicylates, phenoxyacetate herbicides (2,4-dichlorophenoxyacetic acid 2,4-D, mecoprop) and phenobarbitone. Alkalinization of the urine is more important than a large intravenous fluid load. A diuresis of more than 500 ml/hour for long periods may be harmful. Before commencing alkaline diuresis, correct plasma volume depletion, and electrolyte and metabolic abnormalities. Sodium bicarbonate 50 to 100 mEq per litre in 5% dextrose is given IV to ensure that the pH of the urine is between 7.5 to 8.5. Recommended paediatric dose is 1 to 2 mEq of sodium bicarbonate per kg in 15 ml/kg 5% dextrose or 0.45% sodium chloride over 3 to 4 hours.

Acid diuresis

Acid diuresis can be used to enhance the elimination of drugs like amphetamines, quinine, ephedrine and flecainide. This procedure is not recommended because it can cause metabolic acidosis and promote renal failure in the presence of rhabdomyolysis.

Dialysis

Peritoneal dialysis increases the elimination of toxins such as ethylene glycol and methanol, but is much less efficient than haemodialysis. Haemodialysis enhances the elimination of salicylate, lithium, methanol, ethylene glycol and ethanol. Haemoperfusion Haemoperfusion involves the passage of blood through an adsorbent material such as activated charcoal which retains the toxic agents. Haemoperfusion can, within 4 to 6 hours, significantly reduce the body burden of compounds with a low volume of distribution.

ANTIDOTES

If a specific antidote is available, it should be given without delay. For example, in organophosphorus poisoning, oxygen should be given if the patient is cyanosed, followed by atropine and pralidoxime intravenously, if there is any delay in initiating or completing lavage.

SUPPORTIVE THERAPY

Maintenance of a fluid balance chart, ensuring an adequate urine output by giving liberal quantities of oral or IV fluids, monitoring serum electrolytes, anticipating and treating cardiac, hepatic, renal and respiratory failure are some important aspects in the management. Late complications, for example, delayed respiratory failure in organophosphorus poisoning should be anticipated and treated.

ANAPHYLAXIS

Anaphylaxis is a complication of envenomation with insect stings and also ingestion, inhalation and skin contact with some drugs, food and chemicals. The patient may present with hypotension, tachycardia, wheezing due to bronchospasm and an urticarial rash.

He should be made to lie flat with the feet elevated. This potentially fatal condition should be treated immediately with 0.5 to 1.0 ml of 1:1000 adrenaline SC or IM (Paediatric dose: 0.01 ml/kg up to 0.3 ml), followed by hydrocortisone 200 mg IV and promethazine 25 mg IM or IV (Paediatric dose: 6.25 to 12.5 mg). If there is no adequate response, adrenaline may be repeated every 10 minutes. Adrenaline can be given IV in life threatening situations.

For intubated patients, if IV injection is not possible, intrathecal instillation (1 to 3 mg) is recommended. If hypotension is present elevate the foot end of the bed and give 1 to 2 litres of IV fluids. Bronchospasm can be relieved by aminophylline 250 mg administered IV slowly or salbutamol 2.5 mg administered by nebulizer. Give oxygen. In severe cases assisted ventilation may be necessary.

INHALATION

If a patient is exposed to a gaseous poison or toxic fumes, rescuers must ensure that they are not also exposed. After carrying the victim to fresh air, breathing should be checked. An adequate airway must be established and oxygen should be given if there is dyspnoea or cyanosis. If respiration is impaired, oxygen and assisted ventilation may be necessary. Pulmonary oedema is a known complication of some irritant gases and oxygen and assisted ventilation may be useful in the management. If there are no contraindications morphine 10 mg IM and aminophylline 250-500 mg IV slowly may be administered to relieve pulmonary oedema.

EYE CONTACT

If a poison enters the eyes they should be washed thoroughly for 15 minutes with running water. The patient should continuously blink during the procedure. If blurring of vision, pain or redness persists, an ophthalmologist should be consulted. Chemical antidotes should never be used locally on the eye.

SKIN CONTACT

Skin contact with a poison can be harmful if it is absorbed through the intact skin, or if the skin is inflamed or damaged by disease or injury. The contaminated skin should be washed thoroughly with soap and water. If possible, water from a tap, a hose or a shower should be used to wash the body, or water may be poured from a bucket. Clothing should be removed carefully while bathing the skin with a stream of water. Chemical antidotes to neutralize a poison should not be used locally on the skin.

FURTHER MANAGEMENT

Most patients who deliberately ingest poisons may need social, economic or psychiatric support to prevent a fatal recurrence. The attitude of a few medical and paramedical staff to treat these patients with contempt and to discharge them as soon as possible, must change. It is strange that a heavy smoker who suffers from a myocardial infarct or an alcoholic with bleeding oesophageal varices gets sympathy and better care than a young patient with socio-economic problems admitted with deliberate poisoning. Once the patient is fit to be discharged, doctors should carefully assess his or her socio-economic and psychological status. The reason to consume poison must be determined. Doctors or nurses should counsel them adequately before discharge. Patients with psychiatric illnesses like depression and schizophrenia, who have suicidal intent are likely to repeat deliberate poisoning or other methods of suicide. A similar risk exists in patients suffering from incurable illnesses (e.g. cancer, severe disabling rheumatoid arthritis) or chronic painful conditions or when there is a past history of suicidal attempts. They should not be discharged without psychiatric referral. Victims of occupational poisoning need education on safety and preventive aspects.

CHAPTER-2
ENHANCED ELIMINATION

Accidental and intentional poisonings or drug overdoses constitute a significant source of aggregate morbidity, mortality, and health care expenditures. An estimated 2 to 5 million poisonings and drug overdoses occur annually in the United States, although the true incidence is unknown due to underdiagnosis and underreporting.

Management of the poisoned patient begins with a thorough evaluation, recognition that poisoning has occurred identification of the agent involved, assessment of severity, and prediction of toxicity. Therapy involves the provision of supportive care, prevention of poison absorption, and, when appropriate, the administration of antidotes and enhancement of elimination of the poison.

This card will focus on methods to enhance the rate of elimination of poisons following a toxic ingestion. General issues regarding the management of toxic ingestions and specific issues related to decontamination and gastric emptying are discussed separately. (See "General approach to drug intoxications" and see "Decontamination of poisoned patients"). Specific management of common drug overdoses (eg, acetaminophen, aspirin, theophylline, tricyclic antidepressants, methanol) are also discussed fully on the appropriate cards.

Enhanced elimination techniques were employed in approximately 1 percent of poison exposures in 1996. These techniques can accelerate removal of a toxin, but few studies have investigated whether they actually shorten the duration of clinical toxicity and improve clinical outcomes. The main methods of enhancing the elimination of toxins are listed below.

Methods of enhancing chemical elimination of poisons

1. Multiple dose activated charcoal
2. Saline dieresis

3. Urinary ion trapping
4. Extracorporeal methods
   a. Hemodialysis
   b. Hemoperfusion
   c. Hemofiltration
   d. Plasmapheresis
   e. Exchange transfusion
5. Hyperbaric oxygen
6. Chelation therapy
7. Cerebrospinal fluid removal
8. Immunologic therapy (Specific antibody – toxin binding)

General indications for enhanced elimination techniques include:

- Ingestion of a poison whose elimination can be enhanced
- Failure of a patient to respond to maximal supportive care
- The nature of the toxin, the measured toxin concentration, impaired endogenous clearance, and/or comorbid illnesses that predict a complicated course
- The perceived benefit of an intervention outweighs the risks of iatrogenic complications

Multiple-Dose Activated Charcoal

Multiple-dose activated charcoal (MDAC) is the most commonly used method for enhancing the elimination of toxins, but was used in only 0.6 percent of reported toxic ingestions in 1996. MDAC can be beneficial in both the preabsorptive and postabsorptive phases of poisoning. Its effect is most pronounced when large amounts of drugs are ingested or drug dissolution is delayed (eg, ingestion of sustained-release and enteric-coated preparations, or drugs which slow gastrointestinal motility or form concretions).

MDAC can also enhance elimination of absorbed toxins by interrupting enterohepatic and enteroenteric recirculation and by promoting passive diffusion of drugs down a concentration gradient from the intestinal mucosal capillaries into the intraluminal space; this process is referred to as enterocapillary exsorption, reverse absorption, or gastrointestinal (GI) dialysis. The intestinal mucosa serves as a semipermeable dialysis membrane, and the intraluminal binding of free drug by activated charcoal drives further drug diffusion into the lumen. (See "Decontamination of poisoned patients" for a discussion of the use of single-dose activated charcoal).

Efficacy

The effectiveness of MDAC in producing significant acceleration of drug elimination is largely dependent upon the characteristics of the ingested drug. MDAC is most effective in removing drugs with a high charcoal binding capacity, a low intrinsic clearance (ie, a prolonged elimination half-life), a small volume of distribution, low protein-binding, and a nonionized state at physiologic pH (low pKa).

Controlled animal and human volunteer studies and uncontrolled studies in poisoned patients have shown that MDAC is capable of accelerating the elimination of carbamazepine, dapsone, phenobarbital, quinine, and theophylline. In one study involving human volunteers, for example, MDAC significantly reduced the mean elimination half-life of intravenously administered phenobarbital from 110 to 45 hours.

There has been only one randomized, controlled, clinical trial comparing MDAC with single-dose activated charcoal in actual poisoned patients. Ten comatose patients poisoned with oral phenobarbital were randomized to either MDAC or single-dose activated charcoal; although the mean elimination half-life was significantly lower in the group that received MDAC (36 versus 52 hours), the length of time that the patients in each group required mechanical ventilation and hospitalization did not differ. However, the lack of clinical benefit from MDAC in this study may have been related to its small size and limited power to detect significant differences in clinical outcomes.

Indications

Based largely upon experimental models rather than empiric data, MDAC is probably beneficial for a small number of drugs (table 2). For some agents (eg, phenobarbital, theophylline, dapsone, and carbamazepine), MDAC produces clearance rates that approximate those of hemodialysis or hemoperfusion.

Dose

The effectiveness of MDAC is determined by the total dose of activated charcoal administered rather than the dosing interval. Activated charcoal at one, two, and four hour intervals will not differ in efficacy as long as the total dose administered with each dosing regimen is identical. When administering MDAC, an initial dose of 1 g/kg of activated charcoal with sorbitol followed by 0.5 to 1 g/kg of activated charcoal in aqueous suspension every two to four hours is recommended. Only the first dose of activated charcoal should be given with sorbitol.

The charcoal dosing regimen actually delivered is often dictated by patient tolerance. In patients with vomiting or decreased gastrointestinal motility, smaller, more frequent doses or slow continuous nasogastric infusion may be most appropriate because of better tolerance. MDAC should be continued until there is significant clinical improvement or until serum drug concentrations have fallen to nontoxic levels. In general, MDAC is not required for more than 24 hours.

Contraindications

Contraindications to MDAC are identical to those for single-dose activated charcoal. Neither should be employed in patients with gastrointestinal ileus, perforation, or obstruction, or in patients with depressed mental status and an unprotected airway.

Complications

MDAC is generally safe and free from serious side effects when administered appropriately in controlled settings. However, MDAC is associated with a greater risk of complications than single-dose activated charcoal. MDAC has caused intestinal obstruction, infarction, and perforation due to inspissation of charcoal, and the risks of abdominal cramping, diarrhea (with excessive cathartic use), constipation, vomiting, and fatal pulmonary aspiration are greater than when a single dose is used.

Agent for which multiple doses activated charcoal may enhance elimination

Probably effective drugs: carbamazepine, dapsone, phenobarbital, quinine, theophylline

Possibly effective: amitriptyline, cyclosporine, diazepam, digitoxin, digoxin, disopyramide, doxepin, glutethimide, meprobamate, methotrexate, nadolol, nortriptyline, phencyclidine, salicylates, sotalol, and valporate.

Urinary Alkalinization and Forced Diuresis

The urinary excretion of some drugs can be enhanced by increasing the urine output and, more importantly, altering the urine pH. The latter is used to convert the lipid-soluble intact acid (HA) or base (BOH) in the tubular lumen into the charged salt (A- or B+):

$$HA \longleftrightarrow H^+ + A^-$$

$$BOH \longleftrightarrow B^+ + OH^-$$

The charged particle is lipid-insoluble and cannot diffuse back into the extracellular fluid, thereby leading to a marked increase in drug excretion.

For drugs that are weak acids (eg, salicylates, phenobarbital), raising the urine pH to 7.5 to 8.0 will drive the first reaction to the right, producing the desired increase in the concentration of the charged salt (A-). This intervention will not affect the elimination of short-acting barbiturates, which are primarily metabolized in the liver.

On the other hand, urinary acidification (urine pH below 5.5) with ammonium chloride or ascorbic acid will increase the excretion of weak bases by driving the second equation to the right. This will be beneficial with intoxications due to amphetamines, quinidine, or phencyclidine. However, this intervention is not recommended because its efficacy has not been established and toxicity can occur.

Indications and efficacy

Drugs which are likely to respond to urinary alkalinization and forced diuresis usually meet four criteria:

- They are predominantly eliminated unchanged by the kidney
- They are distributed primarily in the extracellular fluid compartment
- They are minimally protein-bound
- They are weak acids with pKa ranging from 3.0 to 7.5

The use of forced saline and alkaline diuresis is limited but may be useful for the drugs listed in Tables 3 and 4 (show table 3 and show table 4). Although forced alkaline diuresis reduces the half-life of certain drugs, it tends to be less effective than MDAC or hemodialysis. As an example, one study of ten volunteers who were administered phenobarbital found that MDAC was superior to urinary alkalinization in enhancing the elimination of the drug, with elimination half-lives of 19, 47, and 148 hours in the MDAC, alkalinization, and control arms, respectively.

The clinical course of phenobarbital-poisoned patients appears to be improved by the use of urinary alkalinization and forced diuresis. One study of 16 phenobarbital-poisoned patients found that the concomitant use of both techniques doubled the rate of elimination and shortened the period of unconsciousness by up to 50 percent compared to patients receiving supportive care only. No comparable data exist showing improved clinical outcomes with the use of urinary alkalinization and forced diuresis in other types of poisoning.

Both forced diuresis and urinary alkalinization also increase salicylate clearance. Urinary alkalinization is the most effective single method short of hemodialysis to enhance salicylate excretion, producing a mean elimination half-life of 5 hours versus half-lives of 8 and 19 hours in patients treated with forced diuresis or standard supportive care, respectively. In contrast to the above findings with phenobarbital, this study did not compare clinical outcomes with those seen using supportive care alone.

Technique —The goal of forced alkaline diuresis is to achieve a urine flow rate greater than 3 mL/kg per hour with a urine pH of 7.5 or higher. Prior to the initiation of therapy, baseline measurements of electrolytes, blood urea nitrogen, serum creatinine, glucose, arterial (or venous) and urinary pH, and serum drug concentrations should be performed. The placement of a bladder catheter to accurately measure urine output is recommended.

Urinary alkalinization is achieved by the intravenous infusion of half-normal saline or five percent dextrose in water to which 50 to 150 meq of sodium bicarbonate has been added to each liter to produce a final solution that is roughly isotonic. Normal saline may be used in situations in which forced diuresis without urinary alkalinization is required.

The initial rate of fluid administration should be dictated by a patient's fluid status and renal function; administration of one to two liters over the first hour is usually appropriate. After initial volume expansion, fluid administration should be titrated to equal urine output. The amount of sodium bicarbonate added to each liter of intravenous fluid should be adjusted according to the urine pH. Increases in serum pH are not clinically important in patients with normal renal function.

The administration of 20 to 40 meq/L of potassium chloride may be required if the plasma potassium concentration falls during urinary alkalinization. Diuresis may be promoted by adding furosemide or an osmotic diuretic such as mannitol. Acetazolamide should not be used to alkalinize the urine because it will lower the serum pH, possibly worsening clinical toxicity. Close monitoring of blood and urine pH, electrolytes, respiratory status, and urine output is important when diuresis and urinary alkalinization procedures are performed.

Contraindications

Forced alkaline diuresis is contraindicated in patients with congestive heart failure, renal failure, or cerebral or pulmonary edema.

Complications

Complications of forced alkaline diuresis include fluid overload, pulmonary edema, cerebral edema, hypernatremia (if no free water is given), hypokalemia, alkalemia (which is a problem only in patients with impaired function who cannot excrete the bicarbonate load) and ionized hypocalcemia (which is generally asymptomatic).

Agent for which forced dieresis may enhance elimination

Barium, bromides, chromium, cisplatin, cyclophosphamide, 5-flurouracil, iodide, thalium, calcium, fluride, meprobamate, potassium, lithium and isoniazid

Agents for which alkaline dieresis may enhance elimination

Chlorpropamide, salicylates, diffunisal, fluoride, methotrexate, barbiturates and sulphonamide

Hemodialysis and Hemoperfusion

Hemodialysis (HD) was performed in only 0.04 percent of all poison exposures reported to the American Association of Poison Control Centers in 1996. However, HD was used in 90 percent of patients in whom an extracorporeal method was employed to enhance poison elimination. During HD, up to 400 mL of blood per minute passes through an extracorporeal circuit in which toxic compounds in blood diffuse through a semipermeable membrane down a concentration gradient into a dialysate. Electrolyte disturbances and metabolic acidosis induced by certain drugs also can be readily corrected with this intervention.

Hemoperfusion refers to the circulation of blood through an extracorporeal circuit containing an adsorbent such as activated charcoal or polystyrene resin. In contrast to HD circuits, hemoperfusion devices contain thin, highly porous membranes and adsorbents that provide a large surface area to directly bind toxins. Clearance rates are higher with hemoperfusion than HD if the adsorbent binds the ingested toxin; the extraction ratio for hemoperfusion approximates 1.0 for some poisons, and drug clearance rates approach the rate of blood flow through the hemoperfusion circuit. Peritoneal dialysis is much less effective than HD or hemoperfusion. It is indicated only when HD or hemoperfusion methods are unavailable, contraindicated, or not possible (eg, neonates).

Efficacy

HD is most useful in removing toxins with the following characteristics:

- Low molecular weight (< 500 daltons)
- Small volume of distribution (< 1 L/kg)
- Low degree of protein-binding
- High water solubility
- Low endogenous clearance (< 4 mL/min per kg)
- High dialysis clearance relative to total body clearance.

The utility of HD and hemoperfusion is limited when most of the drug is stored outside of the extracellular fluid because of high lipid solubility and/or tight tissue binding. These characteristics are present with tricyclic antidepressants, digoxin, and calcium channel blockers.

In conjunction with ethanol- or fomepizole-induced blockade of alcohol dehydrogenase, hemodialysis is effective at accelerating the clearance of methanol and ethylene glycol and their toxic metabolites, correcting metabolic acidosis, and reducing end-organ sequelae and mortality associated with these poisonings. HD also substantially increases the rate of elimination of isopropanol, salicylate, theophylline, and lithium, although data regarding clinical endpoints are sparse.

Drugs which are adsorbed by activated charcoal can be extracted by hemoperfusion. The rate of removal exceeds that achieved with hemodialysis when the toxin is highly protein bound, has a high molecular weight, or is lipid-soluble. With theophylline, for example, the extraction ratio with hemodialysis is approximately 50 percent as compared to 99 percent at the beginning of hemoperfusion (before the cartridge becomes saturated). However, the extraction ratio only reflects the percent removal of drug that is presented to the dialysis membrane or hemoperfusion cartridge; thus, for drugs with prominent tissue stores, these techniques remove only a small fraction of the body load.

Despite high extraction ratios and clearance rates that can be obtained with hemoperfusion for some drugs, these factors do not necessarily predict enhanced clinical effectiveness and a more favorable outcome in poisoned patients. No controlled clinical studies in poisoned patients have been performed to determine if hemoperfusion reduces morbidity or mortality as compared to supportive measures. Evidence of clinical effectiveness for hemoperfusion is based upon favorable pharmacokinetic data, animal studies, anecdotal case reports, case series, and uncontrolled retrospective studies in poisoned patients. Three clinical studies have retrospectively compared hemoperfusion with supportive care for poisoning from a variety of drugs, but do not allow firm conclusions to be drawn regarding the relative efficacy of different management strategies.

Indications

When intoxication has occurred with a drug whose HD clearance is significantly greater than endogenous clearance, the use of HD should be considered if the patient's condition progressively deteriorates or when measured drug concentrations are predictive of a poor outcome without HD. Practically, HD is indicated for a limited number of poisonings.

It is significantly more effective than HD in enhancing the clearance of theophylline but is associated with a higher complication rate and is not available at many medical centers. If hemoperfusion is available, it is preferred over hemodialysis.

Technique

HD and hemoperfusion require central venous access with a double lumen catheter. Acute vascular access for hemodialysis or hemoperfusion is best accomplished with femoral catheters, which can be rapidly and safely inserted. Subclavian catheterization can also be used; however, there is a risk of pneumothorax or hemothorax, and therapy may be delayed because of the necessity for radiologic confirmation of proper catheter placement. The duration of these procedures for poisoned patients is usually four to eight hours but should be governed by the clinical response and serum drug concentrations.

Contraindications

HD and hemoperfusion normally require systemic anticoagulation with heparin, and patients with active hemorrhage or coagulopathy may not be candidates for these procedures. HD and hemoperfusion also may not be possible in hypotensive patients.

Complications

Potential side effects of HD include hypotension, bleeding due to anticoagulation, hypothermia, air embolus, and complications that may result from obtaining central venous access (show table 7). In addition to these complications, HP also poses potential risks of charcoal embolization, hypocalcemia, hypoglycemia, leukopenia (10 percent reduction), and thrombocytopenia (30 percent reduction).

Common agents for which hemodialysis enhance elimination

Barbiturates, bromides, chloralhydrate, alcohols, lithium, procainamide, theophylline, salicylates, heavy metals, trichloroethanol, atenolol and sotolol.

Common agents for which hemoperfusion may enhance elimination

Barbiturates, sedative and hypnotics, phenytoin disopyramide chloramphenicol, paraquat, amanita mushrooms, carbamazepine, valporate, procainamide, caffeine, methotrexate and carbon tetrachloride.

Hemofiltration

There are only limited data available on drug removal by continuous arteriovenous or venovenous hemofiltration. Hemofiltration has been used to enhance elimination of aminoglycosides, vancomycin, and metal chelate complexes, but the technique does not remove highly protein-bound drugs effectively. It may also be of benefit for intoxications with drugs that have a large volume of distribution, tight tissue binding, or slow intercompartmental transfer (such as procainamide).

Blood may be pumped by the patient's own arterial (CAVH) or venous (CVVH) pressure or by a hemodialysis machine entrained in the circuit (CAVHD, CVVHD). Blood that enters the hemofiltration circuit passes through filters (sheet membrane or hollow fiber) with large pores, and an ultrafiltrate forms which drags solutes with molecular weights up to 50,000 daltons (depending upon hemofilter pore size). Cells and solutes larger than the pore size remain in the blood and return to the circulation. In contrast to HD or hemoperfusion, CAVH is driven by the patient's own blood pressure and can be run continuously. The rate of fluid removal, which is equivalent to the plasma clearance of drug, can exceed 100 mL/h; thus, fluid replacement is an essential component of the hemofiltration regimen.

Complications of hemofiltration include clotting of the filter and bleeding due to the requisite use of heparin. Fluid and electrolyte losses from the ultrafiltrate must be replaced continuously.

Exchange Transfusion

Exchange transfusion refers to the removal of a quantity of blood from a poisoned patient and its replacement with an identical quantity of whole blood; the process is usually repeated two to three times. Exchange transfusions are rarely indicated but may be useful in the treatment of massive hemolysis (eg, due to arsine or sodium chlorate poisoning), methemoglobinemia, sulfhemoglobinemia (eg, secondary to hydrogen sulfide exposure), or in neonatal drug toxicity. Complications of the technique include transfusion reactions, ionized hypocalcemia, and hypothermia.

Recommendations

The vast majority of patients who have ingested a poisonous substance or a toxic quantity of a drug can be managed with supportive measures and the administration of a single dose of activated charcoal. However, the severity of the ingestion and pharmacologic properties of the toxin should prompt consideration of techniques to enhance elimination of the poison in approximately 1 percent of cases. Few studies have assessed changes in clinical outcomes when enhanced elimination techniques are employed, and efficacy of these techniques has been judged primarily on the basis of improvement in elimination kinetics. Each technique is associated with potential complications, and the decision to use it should be based upon the drug ingested, the actual and predicted severity of poisoning, the presence of contraindications to the technique, and the effectiveness of alternative methods of treatment.

Despite these concerns, the following general recommendations can be offered:

Multiple dose activated charcoal should be considered following ingestions of carbamazepine, dapsone, phenobarbital, quinine, salicylates, theophylline, and agents that have delayed dissolution characteristics (eg, enteric-coated and sustained-release drugs), or have a large degree of enterohepatic or enteroenteric recirculation. Urinary alkalinization and forced diuresis should be considered after poisonings with agents that are weak acids and eliminated to a large extent in the urine (eg, salicylates, phenobarbital). Hemodialysis should be considered for patients with significant ingestion of alcohols, theophylline, lithium, or salicylates. Hemoperfusion is an alternative to hemodialysis and may result in more rapid clearance of toxins such as theophylline, carbamazepine, valproic acid, or procainamide. Peritoneal dialysis, hemofiltration, and exchange transfusion are rarely indicated in the management of poisoned patients.

CHAPTER-3
TOXICOKINETICS

Introduction

Toxicokinetics (TK) is defined by The International Conference on Harmonization (ICH) as 'the generation of pharmacokinetic data, either as an integral component in the conduct of non-clinical toxicity studies or in specially designed supportive studies, to assess systemic exposure. While developing a molecule as a therapeutic agent researchers consider not only benefit but also risk associated with it. Simply it means if the safety/risk ratio is balanced or safety is more then it will be used as good therapeutic agent. Hence toxicological evaluation got more importance in drug development stages especially in preclinical stage. The need for toxicokinetic data and the extent of exposure assessment in individual toxicity studies should be based on a flexible step-by-step approach and a case-by-case decision making process to provide sufficient information for a risk and safety assessment. Several guidelines have been recommended for the toxicokinetic measurements. These measurement procedures may provide a means of obtaining multiple doses pharmacokinetic data in the test species, avoidance of duplication of studies of such studies when appropriate parameters were monitored; optimum design in gathering the data will reduce the number of animals required (replacement, reduction and refinement{3R}). However this toxicokinetic data focus on the kinetics of a new therapeutic agent under the conditions of the toxicity studies themselves. Dynamic development process of a pharmaceutical product is involves continuous feed-back between non-clinical and clinical studies, no detailed recommendations required for the application of toxicokinetic data to be collected in all studies and scientific judgment should dictate when such data may be useful.

The primary objective of the toxicokinetic studies is to describe the systemic exposure achieved in animals and its relationship to dose level and the time course of the toxicity study. Second, exposure data in animals should be evaluated before human clinical trials. Third, choice of species and treatment regimen used in non clinical studies. Lastly, information on systemic exposure of animals during repeated-dose toxicity studies is essential for the interpretation of study results, to the design of subsequent studies and to the human safety assessment. In addition to all these sex and inter-animal variability also need to be compared, because there are some variations depending up on the species and gender. For example drugs, such as pentobarbital, morphine and methadone, female rats have a much lower liver metabolism than males, resulting in higher plasma levels. A recent survey by the Japanese Pharmaceutical Manufacturers Association compared the results from 102 repeat-dose toxicity studies (ranging from one to 12 months) in mouse, rat, dog and monkey **6**. Sex differences were observed in 41 out of 92 of the studies, primarily consisting of higher exposure in female rats. Second example, species difference, clearance of nicardipine in the plasma of rats is high compared with other species, including humans. Due to hepatobiliary saturation (major metabolic pathway) in dog, proxicromil (anti allergic compound) causes liver toxicity (with elevated plasma levels) but not in the rat or monkey.

Principles involved in toxicokinetics

Quantification and extent of exposure

The exposure might be represented by plasma (serum or blood) concentrations or the AUCs of parent compound and/or metabolite(s) and sometimes by tissue concentrations. Quantification of exposure provides an assessment of the burden on the test species and helps in the interpretation of similarities and differences in toxicity across species, dose groups and sexes. When designing the toxicity studies, the exposure and dose-dependence in humans at therapeutic dose levels (either expected or established), should be considered in order to achieve relevant exposure at various dose levels in the animal toxicity studies. Species differences in the pharmacodynamics of the substance (either qualitative or quantitative) should also be taken into consideration because sometimes it may have other effects. This information may allow better interspecies comparisons than simple dose/body weight (or surface area) comparisons.

Extent of exposure

Systemic exposure should be estimated in an appropriate number of animals and dose groups to provide a basis for risk assessment. Concomitant toxicokinetics may be performed either in all or a representative proportion of the animals used in the main study or in special satellite groups. Both male and female animals are utilized in the main study it is normal to estimate exposure in animals of both sexes unless some justification can be made for not so doing. Toxicokinetic data is not mandatory for studies of different duration if the dosing regimen is essentially unchanged.

Sampling points

In concomitant toxicokinetic studies the time points for collecting body fluids should be as frequent as is necessary, but not as frequent as to interfere with the normal conduct of the study or to cause undue physiological stress to the animals. There are also strict restrictions on blood volume available (no more than 10% of circulating volume can be taken). Sample size is typically 0.25–0.50 ml day−1 in rodents and up to1ml day−1 in non-rodents In each study, justification of number of time points should be made on the basis that they are adequate to estimate exposure. The justification should be based on kinetic data gathered from earlier toxicity studies, from pilot or dose range finding studies, from separate studies in the same animal model or in other models allowing reliable extrapolation. Sampling times vary based on the presence (or lack) of pharmacokinetic data, but are often taken 0.5, 1.0, 2.0, 4.0, 8.0, 12.0 and 24.0 h post-dose, with only the parent drug generally being measured.

Dose level setting

Dose level for toxicity studies is largely regulated by the toxicology findings and the pharmacodynamic responses of the test species. At low dose levels, preferably a no-toxic-effect dose level, the exposure in the animals of any toxicity study should ideally equal or just exceed the maximum expected (or known to be attained) in patients. This ideal is not always achievable and that low doses will often need to be determined by considerations of toxicology; nevertheless, systemic exposure should be determined.

Intermediate dose levels should normally represent an appropriate multiple (or fraction) of the exposure at lower (or higher) dose levels dependent upon the objectives of the toxicity study. The high dose levels in toxicity studies will normally be determined by toxicological considerations. However, the exposure achieved at the dose levels used should be assessed. This toxicokinetic data indicate that absorption of a compound limits exposure to parent compound and/or metabolite(s), the lowest dose level of the substance producing the maximum exposure should be accepted as the top dose level to be used (when no other dose-limiting constraint applies). In non-linear kinetic cases a very careful attention should be paid to the interpretation of toxicological findings in toxicity studies (of all kinds). However, non-linear kinetics should not necessarily result in dose limitations in toxicity studies or invalidate the findings; toxicokinetics can be very helpful in assessing the relationship between dose and exposure in this situation.

Ratifying factors on study to be considered

Earlier we discussed the species and sex differences and their effect on toxicokinetics. There are other factors to be considered in this study is protein binding, tissue uptake, receptor properties and metabolic profile. Systemic exposure may be decreased by protein binding and tissue uptake. In addition, due to the metabolism there will be formation of pharmacological active metabolites, the toxic metabolites and antigenic biotechnology products metabolites.

Route of administration

Pharmacokinetics of a substance is greatly affected by the route of administration. For instance orally administered drugs bioavailability time is more than other routes. If the drug is intended to administer through oral route then oral toxicity should be checked. If any drug administering route is already established and new clinical route of administration is going to establish then it will be necessary to ascertain whether changing the clinical route will significantly reduce the safety margin. In this case focusing on local toxicity is essential. In this comparison of the systemic exposure to the compound and/or its relevant metabolite(s) (AUC and C_{max}) in humans generated by the existing and proposed routes of administration is required.

Metabolite determination

Many of the cases systemic exposure and toxic effect consider on the basis of parent drug concentration. However, there may be circumstances when measurement of metabolite concentrations in plasma or other body fluids is especially important in the conduct of toxicokinetics. They are

- If it is a 'pro-drug' and the delivered metabolite is acknowledged to be the primary active entity.
- If the compound is metabolised to one or more pharmacologically or toxicologically active metabolites which could make a significant contribution to tissue/organ responses.
- For the drugs which are extensively metabolized and the metabolite is only the quantifiable factor.

Statistical evaluation of data

The data should be evaluated statistically which allows assessment of the exposure. Toxicokinetic values are normally calculated as mean SD; statistical evaluation is not usually performed however, because large intra- and inter-individual variation of kinetic parameters may occur and small numbers of animals are involved in generating toxicokinetic data, a high level of precision in terms of statistics is not normally needed. Consideration should be given to the calculation of mean or median values and estimates of variability, but in some cases the data of individual animals may be more important than a refined statistical analysis of group data. If data transformation (e.g. logarithmic) is performed, a rationale should be provided.

Analytical methods

Regulatory authorities expects that analytical methods used to determine plasma concentrations of pharmaceuticals are of adequate sensitivity and precision[1,9]. For evaluation validated analytical methods used and conforms to Good Laboratory Practice (GLP). Analytical methods used in such studies include gas chromatography (although this is rarely used), HPLC (UV or fluorescence), LC, LC–MS, LC-MS-MS, and capillary electrophoresis (again, rarely used, and more for proteins). In addition, the number of sample time-points must be to be frequent enough to estimate exposure[1]. Generally, toxicity studies use a range of time-points and replicates to provide toxicokinetic data as we discussed earlier, although staggered and sparse sampling (to reduce animal numbers) has been reported to give accurate results[10–13]. Results are then analysed using a set curve-prediction package (e.g.WinNonLin from Pharsight; http://www.pharsight.com). For replicate designs, toxicokinetic measurements are taken at similar pre-set timepoints and the mean of the measured values is then taken to provide an estimate of drug exposure. Although staggered designs are less sensitive, they are still used by various pharmaceutical companies for rodent and/or primate studies.

Toxicokinetic studies in Preclinical stage

Safety assessment

Generally safety of a molecule can be performed in in-vivo systems. This step is not included in the guidelines but it is very useful for the researchers to assess the systemic exposure of the molecule and its effect on it. This safety study is integral part in the central nervous system (CNS), cardio vascular system (CVS) and respiratory assessments.

Single dose and rising dose studies

These studies are often performed in a very early phase of drug development before a bioanalytical method has been developed. These studies are usually performed in rodents. Plasma samples may be taken in such studies and stored for later analysis, if necessary; appropriate stability data for the analyte in the matrix sampled would then be required. To answer specific questions raised from the initial single dose study alternatively, additional toxicokinetic studies may be required. Results from single-dose kinetic studies may help in the choice of formulation and in the prediction of rate and duration of exposure during a dosing interval. This may assist in the selection of appropriate dose levels for use in later studies. However, toxicokinetics can be assessed for some drug classes, or in screening studies (e.g.in a series of candidates or when choosing a suitable formulation). Rising-dose studies are performed in non-rodent models. Here, toxicokinetic evaluation takes place at various time-points for each new dose level. Such an evaluation is especially useful if higher-dose emesis occurs as it can reveal whether exposure to the drug still occurred.

Repeated-dose toxicity studies

To give support for phase 1 studies this study is carried out for four weeks in both rodents as well as non-rodents. The treatment regimen (Note 11) and species should be selected whenever possible with regard to pharmacodynamic and pharmacokinetic principles. This may not be achievable for the very first studies, at a time when neither animal nor human pharmacokinetic data are normally available. As we discussed earlier no rigid detailed procedures for the application of toxicokinetics are recommended in regulatory guidance documentation1. Toxicokinetics should be incorporated appropriately into the design of the studies. It may consist of exposure profiling or monitoring (Note 1) at appropriate dose levels at the start and towards the end of the treatment period of the first repeat dose study. The procedure adopted for later studies will depend on the results from the first study and on any changes in the proposed treatment regimen. Monitoring or profiling may be extended, reduced or modified for specific compounds where problems have arisen in the interpretation of earlier toxicity studies. These results give information on exposure, dose proportionality, sex- and species-difference, and potential accumulation and inhibition, and help to support dose-selection for subsequent studies. Performing further repeated dose studies in both rodent and non rodents up to 6-12 months enable estimation of drug and its metabolite(s) kineticparameter assessment as well as long term clinical exposure assessment. Another point to be considered is a few drugs shows tolerance when it is administered repeatedly.

Genotoxicity studies

Two in vitro studies and one in vivo study is essential to support development of drug. In vivo investigations usually use a rodent micronucleus (bone marrow or peripheral erythrocytes) test or chromosome aberration (bone marrow cells) test. These are the well established studies for the genotoxicity evaluation. There is a regulatory expectation to demonstrate exposure to the drug either with toxicity or toxicokinetic data. In rodents, specific toxicokinetic evaluation might not be necessary as it is possible to cross reference with toxicity studies.

Reproduction toxicity studies

Reproduction toxicity measurements are taken in studies of fertility (rat), embryo-foetal development (rat and rabbit) and peri- or post-natal development (rat).

Studies of fertility

Assessment of fertility toxicity has very important, because most of the drugs used in fertility conditions so has to strengthen at that time. Usually this can be done in rats.

In pregnant and lactating animals:

There is a regulatory expectation for toxicokinetic data in pregnant animals, although no specific guidance is given. Data from non-pregnant animals is useful to set dose levels, and the limitation of exposure is usually governed by maternal toxicity. Toxicokinetics may involve exposure assessment of dams, embryos, foetuses or newborn at specified days. Secretion in milk may be assessed to define its role in the exposure of newborns. In some situations, additional studies may be necessary or appropriate in order to study embryo/foetal transfer and secretion in milk. The point at which toxicokinetic evaluation is performed varies among pharmaceutical companies but often takes place in embryo-foetal studies at the beginning and end of gestation in the main study animals themselves. However, it can also occur in preliminary studies or in main studies with satellite animals.

Carcinogenicity studies

Sometimes drugs are used for longtime for curing purposes, this may lead to the toxicity or carcinogenicity. So lifetime studies in the rodent are needed to support the long-term clinical use of pharmaceuticals and non-rodents can also be used. Dose selection is usually determined as the maximum tolerated dose (MTD), which is a 25-fold AUC ratio (rodent to human), or by dose-limiting pharmacodynamic effects, saturation of absorption, or a maximum feasible dose. Selection based on AUC is less common as a 25-fold ratio is often not feasible. Indeed, at the highest dose level, most drugs do not yield AUC values of more than 5–10-fold the human AUC. There is a regulatory expectation for information on systemic exposure to the parent drug and metabolites. It is recommended that monitoring should occur on a few occasions during the study, although it is not essential for monitoring to occur beyond six months. However, pharmaceutical companies use various strategies for such monitoring times (e.g. Weeks 1, 13, 26 and 52, Weeks 1 and 26, or Weeks 26 and 52). It should be noted that, owing to high variability in plasma concentration, toxicokinetic data from aged rats (above one year old) are not useful for estimating exposure. Sampling times depend on available kinetic data but can range from full profile (up to 24 h) to limited time-points which are earlier stated.

Toxicokinetic studies in clinical phases

Regulatory bodies around the world outlining that toxicity studies are necessary to support human Phase I, II and III studies, and product license application is available. The magnitude of the preclinical toxicokinetic evaluation for each clinical phase varies significantly among pharmaceutical companies. For Phase I investigations the company might only generate toxicokinetic data from the four-week repeat-dose toxicitystudies. Full pharmacokinetic profile (including in vitrometabolism studies), and toxicokinetic measurements from four- and 13-week repeat-dose toxicitystudies prior to Phase I is necessary. Toxicity assessments enable the No Observed Effect Level (NOEL) or No Observed Adverse Effect Level (NOAEL) to be established for a potential new drug, based on clinical observations, bodyweight, food consumption, clinical pathology, organ weights, necropsy examination, and histopathology. Toxicokinetic data from either NOEL or NOAEL [and subsequent toxic level(s)] can be used to give guidance to the clinical investigator by providing suitable safe starting and upper doses in the initial single-dose Phase I study. For further clinical studies using multiple dosing, toxicokinetic data from toxicity studies provide information on possible increases or decreases of drug in plasma. Cases where human plasma levels in a Phase I study are higher than in the animal study NOEL or NOAEL values need to consider the effects of different metabolism and plasma protein binding. This might result in the use of a different species in the toxicity study and/or a change of formulation to enable reassessment of safety margins.

Approaches to decrease the animal usage in toxicokinetics

To increase the generation of toxicokinetic data it has to increase the usage of number of rodents through the use of satellite groups. As discussed earlier using of animal's number and sex is restricted as per OECD-417 guidelines even in blood samplings. For this purpose many alternative approaches has generated.

Dried blood spot technology

Recently toxicokinetic studies in dogs by applying dried blood spot technology published and also it is in improvement stages to use this model in rodents (bar). In Pharmaceutical industry this type of methodology for toxicokinetics has been developed and combined with high performance liquid chromatography–mass spectrometry (HPLC–MS/ MS) to collect and analyse very small amounts of biofluids. By this approach, smaller volumes of blood (typically 10–20 lL per sample) used to generate high quality toxicokinetic information than are conventionally required (200 lL for mice and 250 lL for rats). This could enable a significant reduction in rodent numbers for generation of kinetic data as serial samples can be obtained from the same animal. Toxicokinetic samples can be collected from rats already in the toxicological study, while the number of mice in a satellite group can be reduced.

Alternative approaches to animal models

Using alternative models to animal models can reduce the number of animals in confirmative studies. Although this is not basis for the use of the chemical entity in clinical phases, but prior to human usage it should check with the animal studies. For example carcinogenecity studies of chemical entity, alternative model like cell transformation assay used. In cell transformation Syrian hamster embryo (SHE) cells,.Balb/c 3T3 mice cells and C3H/10T1/2 (puripotent stem cells) cells can be used.In these assays, carcinogenicity of test substances is determined by measuring phenotypic changes such as cell morphology, colony growth patterns and cell adhesion induced by chemicals in mammalian cell cultures. The most widely used of these assays are the Syrian hamster embryo (SHE) assay, the low-pH SHE assay, the Balb/c 3T3 assay, and the C3H/10T1/2 assay. The SHE assay is believed to detect early steps of carcinogenesis, and the Balb/c and C3H10 assays detect later carcinogenic changes. This evaluation should constitute effective analytical methods having good accuracy and precision, adequate sampling, drug and metabolite(s) evaluation both in animals and humans (if necessary) and sufficient results evaluation. Toxicokinetic data is important to know the toxic response(s) to that of drug exposure obtained in drug development stages (preclinical) and it is used to set safe dose for clinical use of new drugs and also it is useful in the understanding of differences in responses or sensitivity between individual animals, genders, species or life stages, and supporting the extrapolation of findings in experimental animals to humans. Kinetic data can also support mode-of-action analysis and extrapolation across exposure routes. Now a day toxicokinetics used in other areas also with other areas of pharmacokinetics. Such as toxicokinetic assessments using biomarkers, are used earlier in screening studies, provide data for allometric species scaling, and even play a role in measuring drug levels in non-plasma samples (tissues, urine and bile). Even though toxicokinetic evaluation is only a small part of the process of understanding the fate of a drug, it has a vital part in drug development – a role that proceeds to advance.

CHAPTER-4
PARACETAMOL POISONING

Paracetamol (acetaminophen) is the most widely used over- the-counter analgesic agent in the world and it is the leading pharmaceutical agent in overdose which leads to hospital admissions. Paracetamol is involved in a large proportion of accidental paediatric exposures and deliberate self-poisoning cases, although hepatic failure and death are the uncommon outcomes. In UK, the proportion of the overdoses with paracetamol increased from 14.3% in 1976 to 42% in 1990, and in 1993, 47.8% of all the overdoses which were reported, involved paracetamol or paracetamol-containing drugs. It has also become increasingly common in countries which include Denmark and Australia. In India, the data on paracetamol self poisoning is uncommon and it is insufficient as compared to that of the west. A 10 year retrospective hospital based study reported 0.32% cases of acute paracetamol overdoses due to accidental exposures. The existing treatment recommendations use oral and intravenous N acetyl cysteine to prevent a hepatic injury and to replenish the glutathione stores.

Paracetamol kinetics

Paracetamol is rapidly absorbed from the small intestine. Its peak serum concentrations occur within 1–2 hours for the standard tablet or the capsule forms. 20% of the ingested dose undergoes first-pass metabolism in the gut wall (sulphation), while the rest undergoes hepatic biotransformation.

The mechanism of paracetamol induced hepatotoxicity can be explained in the following steps:

- 5% of the ingested paracetmol is converted by mixed function oxidases in the hepatocytes into a reactive metabolite, N acetyl p benzoquinonimine.

- In therapeutic doses, this reactive metabolite is conjugated with glutathione and its byproducts, mercapturic acid and cysteine are excreted in urine.
- In cases of a paracetamol overdose, the excess amount of the reactive metabolite accumulates, while the glutathione stores diminish.
- A hepatic toxicity ensues if the glutathione stores drop to approximately 30% of their normal amounts.
- The accumulated reactive metabolite forms covalent bonds with the SH groups in the hepatocytes, resulting in hepatic necrosis.

Risk Assessment

The key factors to consider for paracetamol poisoning are:
- The dose and concentration (early).
- The clinical and laboratory features which suggest liver damage (late).
- A history which suggests an increased susceptibility to the toxicity in alcoholics and malnourishment.
- The serum paracetamol levels should be checked to assess the need for *N*-acetylcysteine administration in all the patients with deliberate paracetamol self-poisoning, regardless of the stated dose. A clinical or biochemical evidence of a liver injury may not be apparent for up to 24 hours after the acute paracetamol overdose.

The pathophysiology of hepatic necrosis following an overdose of acetaminophen has been studied extensively. The following are the hypothesized molecular mechanisms that have been put forth, as has been depicted in.

Lab Assessment

In patients who present within 8 hours after ingestion, the evalu- ation of the serum paracetamol and the alanine aminotransfer- ase (ALT) levels should be performed as soon as possible. Other recommended investigations if the patient is brought in after

8 hours are, Prothrombin time/INR, blood urea and creatinine levels, blood glucose levels and estimation of the arterial blood gas concentration.

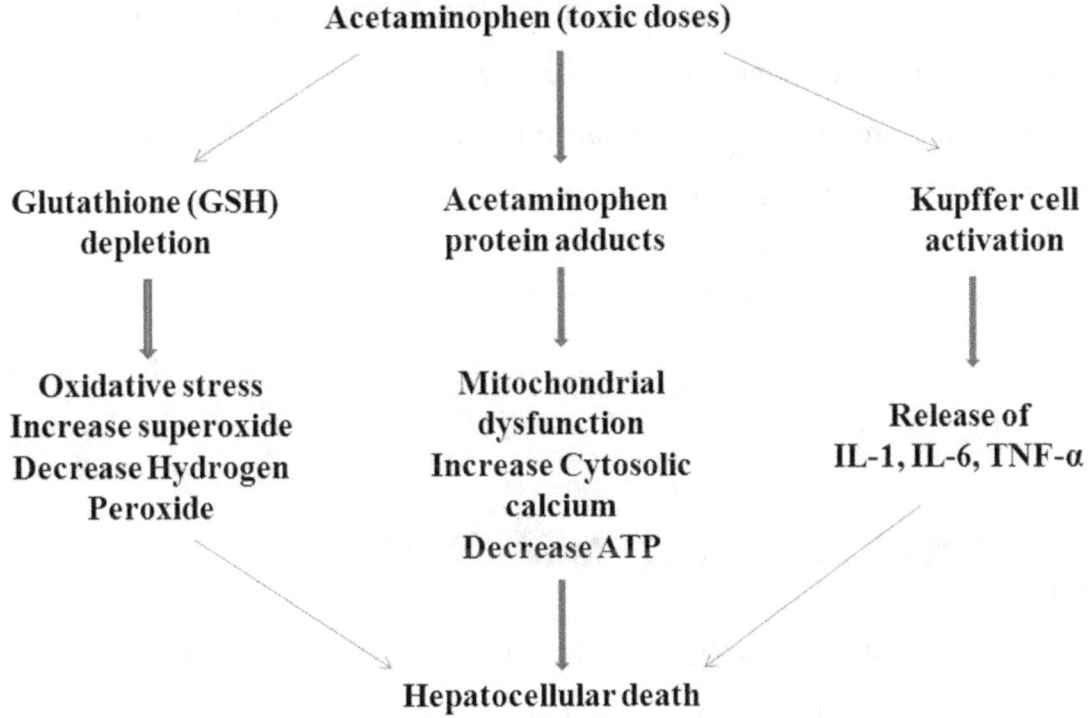

Figure 1: Pathophysiology of paracetamol poisoning

Management of the Paracetamol overdose

A recovery is usually seen with supportive management with *N*-acetylcysteine which is administered in routine doses, although prolonged infusions may be required. The role of haemodialysis has been described in many settings, but the indications for its use have not been clearly outlined.

1. Gastrointestinal decontamination

The administration of activated charcoal within 4 hours of the paracetamol ingestion reduces its further absorption and the further need for N acetyl cysteine administration. The usual dose of activated charcoal in adults is 25-50 gm in 100-200 ml of water (0.5 -1 gm/kg body weight).

2. Antidotes

Since it is the relative scarcity of the SH groups that leads to the hepatotoxicity which is caused by paracetamol, the definitive therapy has been directed towards the measures which are taken to restore it. The first of such agents were cysteine and methionine, which provided encouraging results by replenishing the lost glutathione stores. The use of these agents resulted in dramatic increases in the survival of the patients, but the side effects (flushing, vomiting, etc) which were associated with these therapies led the researchers to seek alternative treatments .These gave way to trials with N-acetyl cysteine, which is now the preferred antidote of choice.

A. Initial infusion
An initial dose of 150mg/kg of N-acetylcysteine diluted in 200mL of 5% glucose and infused over 15 to 60 minutes.
B. Second infusion
Initial infusion is followed by a continuous infusion of 50mg/kg of N-acetylcysteine in 500 mL of 5% glucose over the next 4 hours.
C. Third infusion
Second infusion is followed by a continuous infusion of 100mg/kg of N-acetylcysteine in 1000 mL of 5% glucose over the next 16 hours.
Three-stage N-acetylcysteine infusion

n-Acetylcysteine

Acetylcysteine (also known as *N*-acetylcysteine) prevents the hepatic injury, primarily by restoring hepatic glutathione. It is thought to provide cysteine for the glutathione synthesis and possibly to form an adduct directly with the toxic metabolite of acetaminophen and *N*-acetyl-*p*-benzoquinoneimine and to thus prevent its covalent bonding to the hepatic proteins. In addition, in patients with acetaminophen-induced liver failure, acetylcysteine improves the haemodynamic and oxygen use, it increases the clearance of indocyanine green (a measure of the hepatic clearance), and it decreases the cerebral oedema. The exact mechanism of these effects is not clear, but it may involve scavenging of the free radicals or changes in the hepatic blood flow. When the risk assessment indicates that *N*-acetylcysteine is required, it is administered as a three-stage infusion, with each stage containing different doses, totaling 300mg/kg over 20–21 hours. If hepatic injury is suspected after the three infusion stages, *N*– acetylcysteine is continued at the rate of the last infusion stage (100mg/kg each 16 hours or 150mg/kg/24 hours), until there is a clinical and biochemical evidence of improvement.

Only a small proportion of the patients who present late develop severe hepatotoxicity and fulminant hepatic failure. The clinicians should consult a specialist from the liver unit for advice on the management of patients with liver failure or with signs that indicate a poor prognosis.

Adverse Effects

The most commonly reported adverse effects of intravenous acetylcysteine are anaphylactoid reactions, including rash, pruritus, angio-oedema, bronchospasm, tachycardia, and hypotension. The most severe adverse effects occur with the erroneous dosing of intravenous acetylcysteine in children. These effects include cerebral oedema and hyponatraemia (due to its administration in 5% dextrose). There are rare reports of deaths which were caused due to anaphylactoid reactions.

Recent Advances

IL-1β and other cytokines such as IL-10, macrophage inhibitory protein-2 (MIP-2), and monocyte chemoattractant protein-1 (MCP-1), appear to be involved in hepatocyte repair and in the regulation of proinflammatory cytokines.

Many animal models have used agents that will reverse the oxidative stress mediated hepatic damage which is caused by acetaminophen. Allopurinol, which attenuated acetaminophen protein-adduct formation, mitochondrial dysfunction and oxidant stress, eliminated the hepatocellular nitrotyrosine staining and injury.

The protective effect of deferoxamine against the APAP-induced liver injury may be attributable to the chelation of iron, which can catalyze the generation of active oxygen species in hepatocytes H. Najafzadeh et al., observed that vanadium had a better effect than deferoxamine in the prevention of the hepatotoxicity which was induced by APAP, although the mechanism of its effect was unclear.

The antidote, N- acetylcysteine should be given to all the patients with a serum paracetamol concentration of >200 mg/l. The treatment with *N*-acetylcysteine guarantees a survival if it is administered within 8 hours of the paracetamol ingestion, and the outcome is the same, regardless of when the treatment is given within this 8-hour window. If the antidote is not given, over 60% of the patients with serum paracetamol concentrations above the treatment line may develop serious liver damage, and of these, about 5% will die. Beyond 8–10 hours after the ingestion, the efficacy decreases with an increasing delay in the treatment. The optimal route and the duration of the administration for N-acetylcysteine in the management of the acetaminophen (paracetamol) poisoning are controversial. Recent studies have stated that a shorter hospital stay, patient and doctor convenience, and the concerns over the reduction in the bioavailability of oral N-acetylcysteine by charcoal and vomiting make intravenous N-acetylcysteine preferable for most of the patients with acetaminophen poisoning.

CHAPTER-5
ASPIRIN TOXICITY

Aspirin is one of the oldest medications that remain a part of current practice. Aspirin is a widely prescribed antiplatelet therapy for cardiovascular and cerebrovascular disease. When combined with the fact that aspirin is readily available, aspirin toxicity remains an important clinical problem.

Epidimiology

From 1998-2003, American Association of Poison Control Centers (AAPCC) reported an increase in exposures from 26,972 to 34,828. Accounts for 12.6% of all analgesic-related deaths. Acute overdoses cause morbidity in 16% and mortality in 1% of cases. Chronic overdoses cause morbidity in 30% and 25% mortality in of cases.

Pharmacokinetics

Rapidly absorbed in the stomach. Reach peak levels in 15-60 minutes. 90% bound to albumin in the blood at a dose of 10 mg/dL. 90% metabolized in the liver, 10% unchanged. $T_{1/2}$ = 15-20 minutes. Metabolites and unchanged drug are filtered and secreted by the kidneys

Toxicokinetics

Absorption may be delayed due to bezoar formation or pylorospasm. Peak blood concentrations may be delayed 2-4 hours. 76% bound to albumin at a dose of 40 mg/dL. Increased free drug in the blood. Hepatic enzymes become saturated and elimination follows zero-order kinetics. Functional half-life can be over 20 hours.

Signs & Symptoms of acute toxicity

In salicylate poisoning there generally exists two settings, an acute toxicity and a chronic toxicity. While they both encompass the same signs and symptoms, their presentation can be clinically differentiated. In general, the earliest signs and symptoms of toxicity include nausea, vomiting, diaphoresis, and tinnitus with or without hearing loss. Other CNS presentations include vertigo, hyperventilation, hyperactivity, agitation, delirium, hallucinations which are usually followed by convulsions, lethargy and stupor. A *marked* elevation in temperature is a sign of severe toxicity and typically preterminal condition

Acid-Base

Salicylates clearly stimulate the respiratory center in the brainstem inducing *hyperventilation and respiratory alkalosis.* Salicylates are also weak acids and can replace plasma bicarbonate coupled with impaired renal function (due to ASA toxicity) leading to accumulation of sulfuric and phophoric acids. Salicylates also uncouple oxidative phosphorylation that leads to accumulation of pyruvic and lactic acids and generates large amounts of heat. The end result is a wide anion gap *metabolic acidosis.* Even though the metabolic acidosis occurs at an early stage, the respiratory alkalosis predominates initially. In adults upon presentation to the hospital, ABG reveals the mixed respiratory alkalosis and metabolic acidosis. In children, the initial respiratory alkalosis may be eclipsed by a significant acidosis. This may be due to larger exposure to salicylates per body weight and lesser response of hyperventilation. In a clinical presentation of metabolic *and* respiratory acidosis, suspect salicylate ALI, CNS depression, or severe fatigue.

Glucose Metabolism

Salicylate toxicity appears to produce a discordance between plasma and CSF glucose concentrations. The rate of CSF glucose use far exceeded the supply, even in the presence of normal serum glucose.

Hepatic Effects

There is an increased entry and oxidation of fatty acids into muscle and liver cells. As a result, the concentrations of plasma free fatty acids, phospholipids, and cholesterol decrease. This leads to an increased ketone body formation

Table 1: Differential charecteristics of acute and chronic salicylate poison

	Acute	Chronic
Age	Younger	Older
Biology	Overdose usually intentional	Therapeutic misadventures: Iaterogenic
Diagnosis	Easily recognized	Frequently unrecognized
Other disease states	None	Underlying disorders (Chronic pain condition
Suicidal ideation	typical	No
Clinical differences	Rapid progression of signs	Acute lung injury (ALI), CNS abnormalities
Serum concentration	Marked elevation	Intermediate elevation
Mortality	Un common when recognized, unless ingestion massive	approximately 25%

Neurologic Effects

Toxic doses of salicylates stimulate *then* depress the CNS. Confusion, dizziness, delirium, psychosis, and then ultimately stupor and coma may occur

Otolaryngologic Effects

Large doses (not necessarily toxic) lead to tinnitus, loss of absolute acoustic sensitivity and alterations in perceived sounds

Pulmonary Effects

When ALI or edema are present in a patient with salicylate poisoning the following etiologies must be considered: Aspiration pneumonitis, viral/bacterial infections, postictal and neurogenic ALI and salicylate ALI. Salicylate ALI is a result of adrenergic overactivity producing a shift of blood from systemic to the pulmonary circuit. Leading to pulmonary capillary hypertension and then edema.

GI Effects

Salicylate toxicity results in nausea and vomiting due to local gastric irritation and stimulation of medullary chemoreceptor trigger zones.

Renal Effects

Aspirin use has not been demonstrated to be associated with either chronic nephrotoxicity or ESRD. In adults with glomerulonephritis, cirrhosis, or CKD and in children with CHF, short-term therapeutic doses of aspirin may cause reversible acute renal failure. Possibly due to inhibition of prostaglandins necessary to maintain renal blood flow

Hematologic Effects

Toxicity effects include hypoprothrombinemia and platelet dysfunction

Musculoskeletal Effects

Pure salicylate overdoses can lead to rhabdomyolysis due to uncoupling of oxidative phosphorylation

Diagnosis

Serum salicylate concentrations and concomitant arterial blood pH values can definitively confirm or exclude toxic salicylate levels. Another test that can be used to rule out ASA toxicity is a ferric chloride test to the urine which when positive, shows ASA use. Test is very sensitive to small amounts of salicylates. Trinder spot test. 3 criteria in the 'point of care' setting that can rapidly indicate salicylate poisoning are: Positive urine ketones, Increase in fatty acid metabolism, whole blood glucose and electrolyte determination. Shows decreased bicarbonate and other electrolyte and glucose abnormalities. Shows characteristic acid-base disturbance of salicylate toxicity

Management

Within the initial presentation the use of gastric decontamination (activated charcoal) has shown to reduce the amount of active salicylate by 50-80%. *in vitro* studies suggest that each gram of activated charcoal can absorb approx. 550 mg of ASA. A ratio of 10:1 charcoal to salicylate has the maximum effectiveness. Multiple doses of activated charcoal appears to be superior to single doses although current data is presently insufficient. Fluid replacement is very important in the management of salicylate toxicity. Toxicity can induce major fluid losses through tachypnea, vomiting, hypermetabolic state, and insensible perspiration. The most important management is through urine alkaliniztion. Alkalinization with sodium bicarbonate results in enhanced excretion of ionized acid form of salicylates. Alkalinizing the urine from a pH of 5 to 8 increased renal clearance from 1.3 to 100 ml/min. Do not use acetazolamide due to a concomitant systemic metabolic acidosis and academia. Alkalemia via hyperventilation has many risks and contributes to mortality. Alkalinization can be achieved with a bolus of 1-2 mEq/kg, followed by an IV infusion of 3 ampules of sodium bicarbonate in 1 L of D5% to run at 1.5-2 times maintenance fluid range. Urine pH must be maintained at 7.5-8.0 and hypokalemia must be corrected. Hypokalemia is a common complication due to movement of potassium into cells in exchange for hydrogen ions to compensate for the alkalemia. Calcium should also be measured as decreases are a complication of bicarbonate therapy

Extracorporeal measures (Hemodialysis): These are indicated in specific situations due to the ability to remove salicylates as well as correct fluid, electrolyte, and acid-base disorders. While a more comprehensive therapy, alkalinization of the urine reduces salicylate levels much more rapidly

Indication for hemodialysis

Renal failure, congestive heart failure, acute lung injury, persistent CNS disturbances, progressive deterioration in vital signs. Severe acid-base or electrolyte imbalance, despite appropriate treatment. Hepatic compromise with coagulopathy. Salicylate concentration (acute) > 100 mg/dL (in the absence of the above)

Conclusion

Initial assessment of ASA toxicity is important in the elderly, as the clinical manifestations can mimic other common medical complaints in that age group. Knowledge of drug combinations that include ASA in their formulation is important, as patients may not be aware of its presence.

CHAPTER-6
NONSTEROIDAL ANTI-INFLAMMATORY DRUGS

Non steroidal anti-inflammatory drugs are most frequently prescribed agents in the world. Used by 30-50 million persons daily. 10-20% of persons >65 yo have a current prescription. 40% of elderly Medicaid patients received at least one NSAID prescription.

Aspirin is still the most widely prescribed analgesic-antipyretic and anti-inflammatory agent and is the standard for the comparison and evaluation of the others. Prodigious amounts of ASA are consumed in the US; some estimates place the quantity as high as 10,000 to 20,000 tons annually. Celecoxib (Celebrex- Searle) and Rofecoxib (Vioxx – Merck) are the only two COX-2 inhibitors available in the US. Meloxicam (Mobec- Boehringer-Ingelheim) is in Phase 3 trials in the US; NDA filed Dec 1998. Flosulide is a COX-2 which was discontinued due to renal toxicity (from Novartis). All NSAIDs share the common therapeutic mechanism of inhibiting synthesis of PGs and blocking production of inflammation, pain, or fever. Although the effects of prostacyclins are complex, with one opposing the other, on balance these effects trend toward decreased platelet aggregation and increased vasoconstriction.

Salicylate

Toxic dose: 150-200 mg/kg produce mild toxicity and 300-500 mg/kg produce severe intoxication. Mechanism of toxicity: Uncoupling oxidative phosphorylation and interruption of glucose and fatty acid metabolism. Antidote: Sodium bicarbonate, multiple dose activated charcoal, hemodialysis.

Clinical Manifestations

Acute ingestion: Mixed respiratory alkalemia and metabolic acidosis, coma seizures, hypoglycemia, hyperthermia, pulmonary edema. Death: cardiovascular collapse, CNS failure

Chronic intoxication: Young/ confused elderly, nonspecific presentation: confusion, dehydration, metabolic acidosis. High mortality (up to 25%), lower serum levels

Death: Cerebral and pulmonary edema

Anaphylaxis: Clinical Scenario: 25% of adult asthmatics with *nasal polyps or chronic urticaria* manifest an acute asthmatic attack minutes after NSAID exposure. 2.5% cross-reactivity noted in patients allergic to tartrazine dyes. Mechanism: increased production of LTC_4, LTD_4, LTE_4

NSAID acute toxicity

Clinical manifestations: N/V, metabolic acidosis, CNS and respiratory depression, acute renal failure, aseptic meningitis, hallucinations. Toxic Dose: Generally large doses >6 gms. Villains: Phenylbutazone and mefenamic acid are considered the most toxic due to their ability to provoke all the above symptoms, as well as seizures. Mechanism of toxicity poorly understood. Most experience with Ibuprofen. 73 million prescriptions are written annually for NSAIDs. Reported fatalities are 6-7 times greater for salicylates and APAP than for NSAIDS.

Prevalence of gastric and duodenal ulcers is 9-22%. Bleeding, perforation, or obstruction with 1/10 NSAID-induced peptic ulcer. GI bleeding with 35% of all peptic ulcer complications. Most common serious ADE in US with 10,000-20,000 deaths/year. 200,000-4000, 000 hospitalizations each year in US with > $4 billion health care cost.

Mechanism of GI toxicity

Dual-injury hypothesis of Schoen and Vender, NSAIDs have direct toxic effects on the gastroduodenal mucosa (solid arrows) and indirect effects through active hepatic metabolites and decreases in mucosal prostaglandins (broken arrows). Hepatic metabolites are excreted into the bile and subsequently into the duodenum, where they cause mucosal damage to the stomach by duodenogastric reflux and mucosal damage to the small intestine by ante grade passage through the GI tract.

Table 2: Salicylate toxicity

Factor	Clinical Implications
Elderly	Beware of chronic salicylism; high mortality, difficult diagnosis
Delayed absorption	Perform serial salicylate levels prior to medically clearing
Academia	Redistributes drug to vital organs and enhances toxicity; delays clearance; alkalinize serum/ urine
Hypoxia	Beware of pulmonary edema
Severe overdose	Administer appropriate antidotes and consider dialysis
Anaphylaxis	Recognize association

Table 3: Salicylate toxicity

Factor	Clinical Implications
Nausea and vomiting	Beware of GI Bleed
CNS depression/ seizure	Uncommon, search for other causes
Acidosis	Indicates massive ingestion
Dehydration	Beware of acute papillary necrosis
Known large ingestion	Check renal function

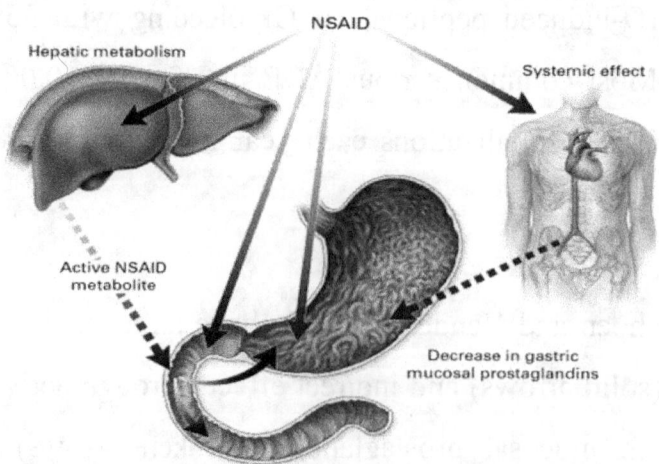

Figure 2: Effect of NSAID

CHAPTER-7

TRICYCLIC ANTIDEPRESSANTS

Epidemiology

More than 18,000 exposures over past decade with tricyclic antidepressants. Incident of toxicity with 60% likely intentional. Increase acute exposure due to narrow therapeutic index of drugs. More drug-related deaths than any other prescription medication. These drugs used for depression, OCD, chronic pain and migraine prophylaxis.

Pathophysiology

All TCA's structurally similar. Cyclobenzaprine structurally similar to TCA's. Many have active metabolites. Which produce multiple toxicologic effects like antihistaminic, anticholinergic, α adrenergic blockade and sodium channel blockage.

Acute toxicity

Acute exposure of TCA is life threatening. Toxic effect started with ingestions greater than 10mg/kg in adults. Pediatrics more susceptible to antimuscarinic effects. Manifest toxic symptoms within 6 hours of ingestion. Coingestion with other cardiac or CNS depressants may be high risk for TCA toxicity. Prior heart disease worsening the symptom of toxicity.

Geriatrics patients are most fatalities ingest more than 1 gram. Fatalities occur in initial hours usually before arrival to hospital

Desipramine

Most potent Na-channel blocker. Twice the fatality rate of other TCA's. May precipitate cardiotoxicity without significant antimuscarinic symptoms. Serious toxicity rarely occurs if <300ng/mL. Most fatalities have levels >1000ng/mL.

Clinical toxicity often does not correlate with serum levels. Urine qualitative may help to rule out TCA toxicity in unknown ingestion

Clinical Features

Varies from mild antimuscarinic to severe cardiovascular collapse. Up to 70% will have coingestants. May have rapid progression of coma and cardiovascular collapse. Mild/Moderate toxicity like drowsiness, confusion, slurred speech, ataxia, dry skin/mucous membranes, tachycardia, urinary retention, myoclonus, hyperreflexia, hypertension. Severe toxicity like coma, conduction delays, hypotension, respiratory depression, ventricular tachycardia, seizures, status occasionally, pulmonary edema, High degree atrio ventricular block.

Treatments

GI Decontamination

DO NOT use syrup of ipecac. Gastric lavage if performed within first few hours after toxic ingestion. Performed lying flat in left lateral decubitus position. Obtunded patients need intubation prior to lavage. Activated Charcoal 1gm/kg PO/NG

Sodium Bicarbonate Therapy

Indications for sodium bicarbonate therapy is widened QRS >100ms, refractory hypotension, terminal R >3mm in aVR, ventricular dysrhythmias. Treatment with sodium bicarbonate improves conduction, contractility and suppresses ventricular ectopy. Dosage initially 1-2 mEq/kg IV bolus until patient improvement or blood pH of 7.50-7.55 by continuous infusion. further adjustments based on blood pH. Hypokalemia is an expected complication. Supplementation usually required

Altered Mental Status

Usually soon after a toxic overdose due to histamine, muscarinic and α-receptor blockade. Administer "DONT" therapy for potentially reversible causes. Consider occult head or neck trauma. Flumazenil and physostigmine contraindicated due to increased risk of seizures

Seizures

Most occur within 3 hours of ingestion. Usually single, but may be multiple in up to 30%. May develop status epilepticus with maprotiline and amoxapine. Benzodiazepines are drug of choice. 2nd line is phenobarbitol 15mg/kg with minor side effects like hypotension, respiratory depression.

CHAPTER-8

ACUTE POISONING OF OPIATES

Opiates work in the brain at specific "opiate receptors". There are several types of opiate receptors but the main receptor is called "Mu". Binding can cause full stimulation or effect at the receptor (agonist), or a partial effect (partial agonist) or block the effect of the receptor (antagonist). Comparison of the amino acid sequences of the cloned mouse deta and kapa and rat mu receptor. The blue indicated common amino acid sequences. G protein couple receptors.

Opiate Receptors And Activation Effect

Mu_1 (μ_1)	analgesia, euphoria
Mu_2 (μ_2)	constipation, respiratory depression
Kappa	spinal analgesia, dysphoria
Delta	analgesia thru the endorphin, enkephalin and dynorphin system

Effect Of Common Opiates At Mu Receptor

Heroin, morphine, methadone	Agonist
Buprenorphine	Partial Agonist
Naltrexone, Nalmefene	Antagonists

Heroin use - urine toxicology can show: Free morphine, morphine glucuronide, free codeine, 6 – Monoacetylmorphine (6 – MAM). This metabolite, or breakdown product, is only seemed with heroin use and no other opiate. It has a very short half – life* and is difficult to detect after heroin use. Half – life is the time it takes to break down 50% of the amount of the consumed substance by the body's metabolic processes.

After IV injection

Warm skin rush. Pruritis (severe itchiness), especially with morphine use which releases histamines. Pleasure, relaxation and satisfaction in 45 seconds

Intoxication or withdrawal?

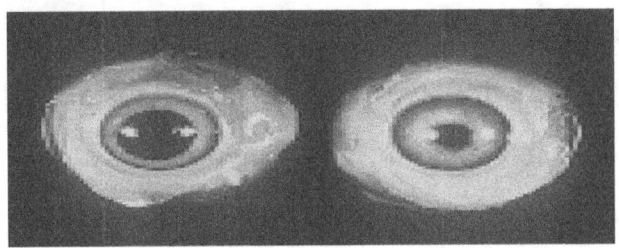

Always look at the pupils; the pupil size can give very good clinical information.

Most common effect

Miosis (small pupils; except with Demerol use which causes paralysis of the ciliary body and pupils dilate). Nodding, hypotension, depressed respiration, bradycardia (slow heart rhythm), euphoria, floating feeling

classic triad seen in overdose

Miosis, coma, respiratory depression, pulmonary edema, seizures.

Opiate Overdose Treatment

Narcan - 0.4 mg IV push, if no response, then 2 mg IV push every 2 - 3 minutes until a total dose of 10 mg is given or a response.

Opiate withdrawal

In general, opiate withdrawal signs and symptoms are the same for all opiates; what differs is the time of onset and the length and intensity of withdrawal. The withdrawal is divided into early, middle and late phases to show the progression of symptoms when the patient is not treated.

Opiate withdrawal symptoms are three phases like early phase, middle phase and late phase

Early phase

Lacrimation (eyes water), yawning, rhino rhea (runny nose), sweating

Middle Phase

Restless sleep, dilated pupils (mydriasis), anorexia, gooseflesh, irritability and tremor

Late phase

Increase in all previous signs and symptoms, increase in heart rate, increase in blood pressure, nausea and vomiting, diarrhea, abdominal cramps, labile mood, depression, muscle spasm, weakness, bone pain.

Opiate Withdrawal Treatment

Can be inpatient or ambulatory detox. Involves the use of medication to damper the increased signs of opiate withdrawal. Clonidine, an antihypertensive medication, has been used in many older protocols. Buprenorphine has recently been approved for use by authorized physicians. Methadone can be used if a detoxification program has the proper approvals. When treating withdrawal medications should always be considered an important component of the treatment. Examples include: Vistaril for mild to moderate anxiety. Oxazepam (15-30 mg q 6 hours) or other benzodiazepine if severe anxiety. Motrin for muscle and joint aches. Tigan for nausea. Kaopectate for diarrhea. Bentyl for abdominal cramps.

Methadone blood levels are available to see if the patient has a therapeutic level. P= peak level where blood is drawn 2 hours after the methadone is given. T = trough where the blood is drawn immediately before the methadone is given (both on the same day). Increase methadone dose if P/T ratio < 2.5 and trough less than 200. Maintain dose of methadone if trough 200- 480 with P/T< 2.5. Decrease dose of methadone if trough > 480 and P/T <2.5. Split dose of methadone if P/T>2.5. Split and increase dose of methadone if trough<200 and P/T>2.5. Split and decrease dose of methadone if peak >960 and P/T>2.5. (If split give 100% in am and 50% pm first day, then 50% twice a day for the next days). 1 - alpha - acetylmethadol acetate Long-acting, orally active analog of methadone, originally approved for use by the FDA in 1993 but is not being manufactured at present by Roxanne Laboratories, Inc. LAAM dose is 1.2 – 1.3 the dose of methadone.

Naltrexone

Highly motivated individuals get benefit from naltrexone. Former opiate-dependent individuals who are employed and socially functioning. Those recently detoxed from methadone/buprenorphine maintenance. Those who are leaving residential treatment settings. Those who sporadically use opiates but are not on methadone/buprenorphine maintenance. Those not eligible for methadone/buprenorphine maintenance. Those in a long waiting period for methadone/buprenorphine maintenance. Those wishing to prevent relapse. Adolescents not wishing to go on methadone/buprenorphine maintenance. Healthcare professionals not wishing to go on methadone/buprenorphine maintenance. For opiate-dependent patients dosing must wait 5 – 7 days after last use of a short-acting opiate (heroin) or 7 – 10 days after a long-acting opiate to prevent withdrawal. Can perform a narcan challenge test* to see if withdrawal can be induced, thus not safe to start naltrexone yet. Should always have a negative urine drug screen for opiates before starting. Start with 25 mg first day, then 50 mg per day thereafter. Can dose for 3 times a week (100mg – 100mg – 150 mg on Monday, Wednesday and Friday)

Complecations

Many of the complications of opiate use are due to the route of use and the lifestyle of the user, not the drug.

Neurologic: Toxic amblyopia (optic nerve pathology), mononeurpathy (dysfunction of a single nerve), polyneuropathy (dysfunction of several nerves), meningitis, brain abscess

Dermatologic: Abscess, tracks, lymphangitis (swelling and dysfunction of the lymph system)

Pulmonary: Aspiration, pneumonia, lung abscess, pulmonary emboli (clots going to the lung), pulmonary fibrosis (scarring of the lung), Noncardiogenic pulmonary edema (lung fills with fluid not as a result of heart dysfunction)

Hepatic: Hepatitis B,C,D,G

Infections: Endocarditis

CHAPTER-9

BARBITURATE AND BENZODIAZEPINE POISON

Barbiturates

Depressants of the central nervous system (CNS) that impair or reduce the activity of the brain by acting as a Gamma Amino Butyric Acid (GABA) potentiators. As an unfortunate fallout, barbiturate overdose became fairly common so much so that it constituted one of the leading causes for self-induced mortality (suicidal or accidental ingestion) in the western countries, and to a great extent in india also. The increase in the abuse of barbiturates may be due to the popularity of stimulating drugs such as cocaine and methamphetamines. The barbiturates ("downers") counteract the excitement and alertness obtained from the stimulating drugs.

Classification

- **Long acting**: (duration of action 6-12 hrs): barbitone, phenobarbitone, mephobarbitone
- **Intermediate acting**: (duration of action 3-6 hrs): Amylobarbitone, butobarbitone, aprobarbitone, amobarbitone
- **Short acting**: (duration of action < 3 hrs): Pentobarbitone, secobarbitone, hexobarbitone
- **Ultra-short acting**: (duration of action < 15-20 min): Thiopentone

Sign and Symptoms of acute poisoning

1st Stage it cause excitement, talkativeness, hallucination, confusion and Lack of Co-ordination. 2nd Stage: Unconsciousness. Pupil fixed. Respiration slow. Reflexes are diminished. 3rd Stage: Coma. At lower doses, a reduction in restlessness and emotional tension occurs. At increasingly higher doses, sedation is followed by increasing levels of anesthesia and eventually death. In large overdose, there is a CNS depression ranging from lethargy to coma , hypotension, pulmonary edema. It also causes hypothermia and ventilatory depression, pupils are usually constricted. With phenobarbitol signs of toxicity usually appear serum concentration exceeds over 4mg/dl. Over dose of barbiturate cause death due to respiratory failure with following symptoms like pupil dilated, pulse weak and rapid, reflex disappear and skin cold and cyanotic.

Chronic poisoning results from regular intake of large doses and is mainly characterised by somnolence , confusion , slurred speech, etc. abrupt withdrawal can provoke an **abstinence syndrome** comprising anxiety, agitation , confusion , tremor, ataxia, convulsions , delusions and hallucination. There is also frequently, insomnia and vomiting.

Treatment

Respiratory depression must be treated with oxygen administration, intubation and assisted ventilation. IV fluids (Ringer's lactate in hypotensive cases). Stomach wash (can be done upto 8 hours post-ingestion). Activated charcoal (1 gm/kg), followed by a suitable cathartic. Administration diuretics are required to enhance the elimination. Forced alkaline diuresis (not very useful in intermediate and short acting barbiturates). haemodiallysis (for long acting) and haemoperfusion (for intermediate and short acting barbiturates). For chronic poisoning, gradual withdrawal must be undertaken together with symptomatic management of adverse manifestations.

Benzodiazepine

Benzodiazepine was introduced in 1960. It produces sedative hypnotic depending on dose and compound. It also called minor tranquillizers. Benzodiazepine was most widely used in human less in veterinary practice. These are more effective and safe as compared to barbiturates with high therapeutic index. These are among the most commonly prescribed drugs today in india , and are mainly used to relieve anxiety. Fortunately, they are relatively safe drugs and rarely cause toxicity.

Acute poisoning effect

Sedation and somnolence, diplopia, dysarthria, ataxia and amnesia

Chronic poisoning effect

Long term, high dose therapy (e.g.30 to 40 mg of diazepam daily), may produce withdrawal reaction when the drug is abruptly stopped. Symptoms : anxiety, insomnia, headache, muscle spasm, anorexia, vomiting, tremor, weakness, etc. rarely, there may be convulsions and psychiatric disturbance.

Treatment

- Decontamination, (stomach wash or emesis, medicinal charcoal and cathartics).
- Doxapram (100mg, IV) and physostigmine have been suggested as antidotes, but serious side effects limit their use.

Flumazenil

Flumazenil (Romazicon) is a specific drug that acts by competitively blocking benzodiazepine receptors thereby preventing benzodiazepine-receptor interactions. Flumazenil's half-life is shorter than those exhibited by benzodiazepines and as a result it would be possible to reverse benzodiazepines-mediated respiratory depression only to have it reoccur upon flumazenil's elimination. Consequently either repetitive flumazenil dosing or continues infusion (0.5-1 ug/kg/min.) may be required to ensure sustained recovery from benzodiazepines mediated effects. Flumazenil reversal tends to have a greater effect on respiratory depression and sedation than on benzodiazepine amnestic properties.

CHAPTER-10

ALCOHOL TOXICITY

Simple alcohols (above) are used widely in industry as solvents, antifreeze or plasticizers. Have very limited use in medicine, as *antiseptics* (dehydrate cells), *antipyretics* (cause evaporative cooling), as *vehicles / solvents* in drug solutions (ethanol / propylene glycol) or as an *antidote* for methanol, isopropanol or ethylene glycol intoxication (ethanol). Absolute ethanol injected into sensory ganglia causes *neurolysis* (destruction) to stop the relay of intractable pain in terminal diseases like cancer.

Ethanol as a beverage component (commonly called "alcohol") along with nicotine and caffeine are essentially the only "recreational" drugs widely legal for lay use by Western society. Alcoholic beverages are major economic commodities, generating substantial tax revenues.

An estimated 90% in USA have consumed alcohol at least once. 10% have "alcohol-related problems" like dysfunctional social relationships with family, friends or co-workers, increased risk and consequences of accidents or crime and a diminished level of health due to ethanol toxicity.

After Oral Ethanol Consumption

Ethanol Absorption is by Passive Diffusion

Ethanol easily crosses membranes like water and is totally absorbed from stomach / upper small intestines. Absorption is slower from the more muscular stomach (~25%) than from the small intestines (~75%). Food or high ethanol concentrations in the stomach, slow absorption, delaying gastric emptying into small intestine where absorption is fastest.

Ethanol Distributes in Total Body Water

Volume of distribution = ~ 0.7 L/kg and **reflects tissue water content**. For example, urine has more water than blood and thus a greater ethanol concentration relative to blood. There are **no tissue barriers to ethanol!**

Target Proteins for Ethanol = $GABA_A$ / NMDA receptors & L-type voltage-gated Ca^{2+} channels. Other targets are known, but their role in intoxication is unclear.

Increases $GABA_A$ Receptor Cl^- Channel Activity, enhancing inhibitory tone allosterically (like benzodiazepines, barbiturates, propofol and halothane, etc.). Interacts at a specific hydrophobic binding pocket between transmembrane regions 2 & 3 of $GABA_A$ receptor subunits to allosterically increase GABA affinity for the receptor and enhance synaptic inhibition. This action likely contributes to anxiolytic, sedative, motor impairing and anesthetic actions of ethanol.

Inhibits NMDARs (like ketamine) reducing CNS excitability. This action likely **contributes to blackouts** during heavy binge drinking. **Inhibits L-type Ca^{2+} (LEFT)** Activation of NMDA receptor Na^+ **channels** reducing CNS excitability. These channels play a major role in vasopressin release. When they are blocked by ethanol consumption, vasopressin release decreases, causing reduced water resorption in the kidney and **alcohol-induced diuresis**. Brain Ca^{2+} channels up-regulate with physical dependence playing a **role in ethanol withdrawal hyper excitability**.

Signs / Symptoms of Ethanol Intoxication

Ethanol consistently impairs behavior (judgment, memory, reaction time, etc.) and **motor coordination** in dose-dependent manner correlating with blood alcohol concentration (BAC).

Respiration depression causes death in acute alcohol overdose by blunting respiratory drive from diminished medullary sensitivity to CO_2. Ethanol overdose deaths are surprisingly common! **Increased sexual desire** but **impaired sexual performance. Loss of thermoregulation** so body temperature follows environmental temperature (e.g., **poikilothermia**), leading to either *hyperthermia* or *hypothermia*, both of which can be life-threatening! **Vasomotor center impairment**

(brain stem) causes **hypotension** and **increased peripheral vasodilation** which further increases risk of hypo- or hyperthermia (gives a false sense of warmth).

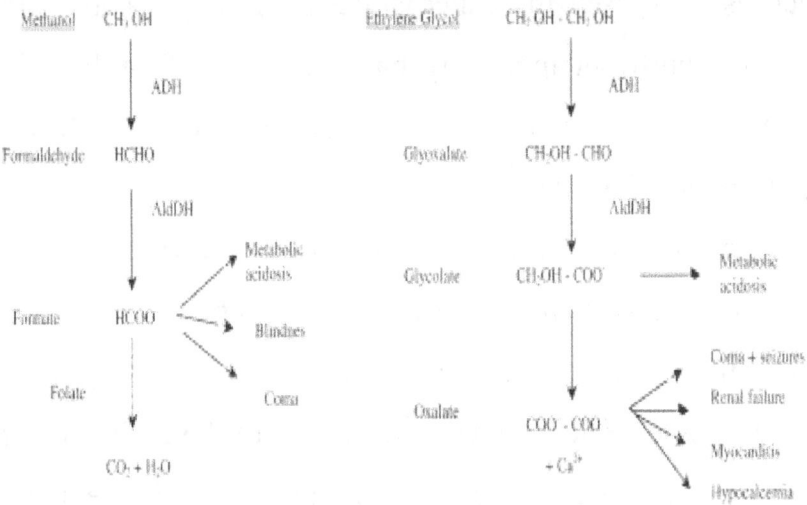

Figure 3: Pathogenesis of methanol and ethylene glycol poisonings.

Impaired neuroendocrine function

Inhibits arginine vasopressin (AVP, antidiuretic hormone) secretion causing diuresis / dehydration. **Blocks oxytocin** secretion to diminish lactation / uterine contractions. **Stimulates secretion of adrenocorticotrophic hormone** (ACTH) and **epinephrine** similar to stress.

Treatment

Provide proper supportive care to limit the absorption of alcohol to avoid systemic toxicity. Administration of specific antidote like fomepizole and disulfiram may prevent the toxic metabolite conversion of alcohol.

Administration of ethanol may prevent the metabolic conversion of methanol and ethylene glycol. The dose of ethanol is 80-130 mg/kg/hour depending on how fast a patient metabolizes and needs to be increased to 250 mg/kg/hour or higher during dialysis.

Sodium bi carbonate and folate administration is adjunct for methanol poisoning. Both enhancing the elimination during hemodialysis process.

Fomepizole is a safe and effective antidote

Fomepizole [4-methylpyrazole (4MP)] is a potent inhibitor of ADH with limited toxicity. It has been successfully used in France since 1981 in EG and methanol poisonings. No lethality or significant morbidity has occurred with either alcohol when patients were treated before significant toxic metabolism occurred; all patients recovered from their poisonings. Two recent U.S. multi-center prospective clinical trials confirmed fomepizole's efficacy. Rapid resolution of acidosis accompanied clinical improvement, with no new symptoms of poisoning after the initiation of therapy. Renal injury did not occur if fomepizole was administered early in EG intoxication. Treatment with fomepizole resulted in alteration of the toxicokinetics of both EG and methanol, with a prolongation of their elimination and a reduction in glycolate and formate formation.

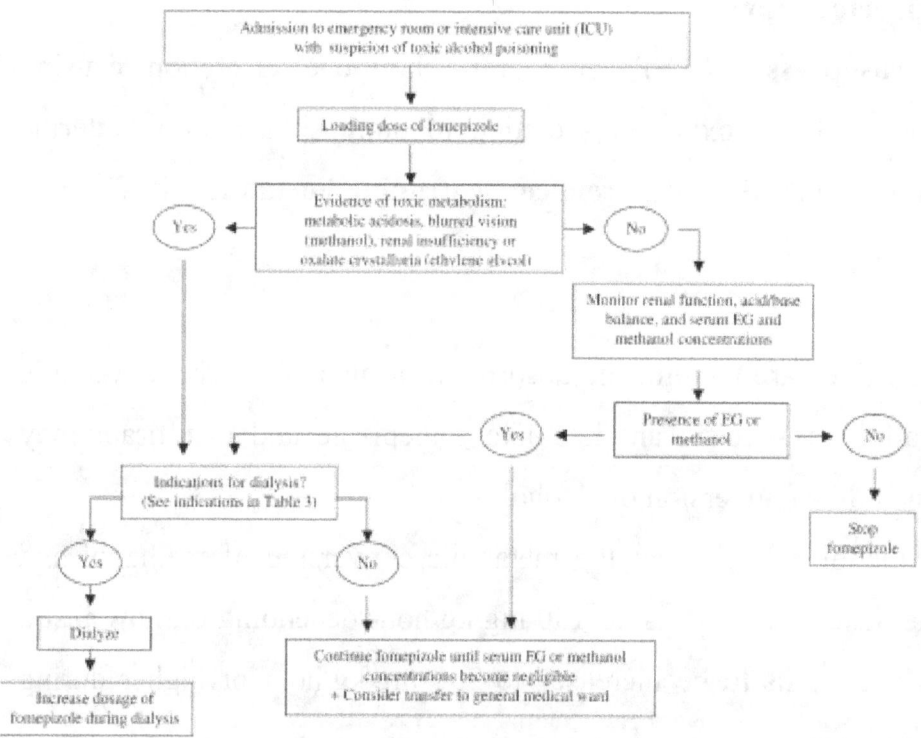

Figure 4: Proposed algorithm for treatment of EG and methanol-poisoned patients

Disulfiram (Antibuse®) - **inhibits aldehyde dehydrogenase** causing acetaldehyde to accumulate triggering "**Alcohol/Antibuse® reaction**" marked by flushing of the skin, dysphoria, nausea, autonomic hyperactivity and headache. Alcoholics *trying to stop drinking* (see Treating Alcohol Dependence, above) must agree to take disulfiram after first experiencing its aversive/dysphoric effects to be better able to resist alcohol drinking. **Disulfiram-like side-effects** also occur with **cephalosporins, metronidazol** or **tolbutamide.** Some racial groups have a *deficiency in aldehyde dehydrogenase* that appears to be protective against alcohol abuse or alcoholism, by elevating acetaldehyde to dysphoric levels that make drinking unpleasant or less reinforcing.

CHAPTER-11

RADIATION POISONING

Radiation Sickness is a form of damage to organic tissue due to excessive exposure to ionizing radiation. Acute Radiation Syndrome (ARS) (sometimes known as radiation toxicity or radiation sickness) is an acute illness caused by irradiation of the entire body (or most of the body) by a high dose of penetrating radiation in a very short period of time (usually a matter of minutes).

Stages of Radiation Sickness

- Prodromal stage (N-V-D stage) – The classic symptoms for this stage are nausea, vomiting and diarrhea that occur from minutes to days following exposure. The symptoms may last (episodically) for minutes up to several days.
- Latent stage – In this stage the patient looks and feels generally healthy for a few hours or even up to a few weeks.
- Manifest illness stage – In this stage the symptoms depend on the specific syndrome (see Appendix) and last from hours up to several months.
- Recovery or death – Most patients who do not recover will die within several months of exposure. The recovery process lasts from several weeks up to two years.

Radiation Sickness sometimes known as Acute Radiation Syndromes has three classes. Table 1 shows the different aspects of radiation syndrome

- Hematopoietic Syndrome
- Gastrointestinal Syndrome
- Cardiovascular/ Central Nervous System Syndrome

Table 4: Various aspects of radiation syndrome

Syndrome	Dose[1]	Prodromal Stage	Latent Stage	Manifest Illness Stage	Recovery
Bone Marrow	0.7 – 10 Gy (70 –1000 rads) (mild symptoms may occur as low as 0.3 Gy or 30 rads)	• anorexia, nausea and vomiting • occurs 1 hour to 2 days after exposure • lasts for minutes to days	• stem cells in bone marrow are dying, though patient may appear and feel well • lasts 1 to 6 weeks	• drop in all blood cell counts for several weeks • anorexia, fever, malaise • primary cause of death is infection and hemorrhage • survival decreases with increasing dose • most deaths occur within a few months after exposure	• in most cases, bone marrow cells will begin to repopulate the marrow • there should be full recovery for a large percentage of individuals from a few weeks up to two years after exposure • death may occur in some individuals at 1.2 Gy (120 rads) • the $LD_{50/60}$[2] is about 2.5 to 5 Gy (250 to 500 rads)
Gastrointestinal (GI)	10 – 100 Gy (1000 – 10,000 rads) (some symptoms may occur as low as 6 Gy or 600 rads)	• anorexia, severe nausea, vomiting, cramps and diarrhea • occurs within a few hours after exposure • lasts about 2 days	• stem cells in bone marrow and cells lining GI tract are dying, though patient may appear and feel well • lasts less than 1 week	• malaise, anorexia, severe diarrhea, fever, dehydration, electrolyte imbalance • death is due to infection, dehydration and electrolyte imbalance • death occurs within 2 weeks of exposure	• the LD_{100}[3] is about 10 Gy (1000 rads)
Cardiovascular (CV)/ Central Nervous System (CNS)	> 50 Gy (5000 rads) (some symptoms may occur as low as 20 Gy or 2000 rads)	• extreme nervousness, confusion; severe nausea, vomiting, and watery diarrhea; loss of consciousness; burning sensations of the skin • occurs within minutes of exposure • lasts for minutes to hours	• patient may return to partial functionality • may last for hours but often is less	• return of watery diarrhea, convulsions, coma • begins 5 to 6 hours after exposure • death within 3 days of exposure	• no recovery

Hematopoietic syndrome – the full syndrome will usually occur with a dose between 0.7 and 10 Gy (70 – 1000 rads) though mild symptoms may occur as low as 0.3 Gy or 30 rads. The survival rate of patients with this syndrome decreases with increasing dose. The primary cause of death is the destruction of the bone marrow, resulting in infection and hemorrhage.

Gastrointestinal (GI) syndrome – the full syndrome will usually occur with a dose between 10 and 100 Gy (1000 – 10,000 rads) though some symptoms may occur as low as 6 Gy or 600 rads Survival is extremely unlikely with this syndrome. Destructive and irreparable changes in the GI tract and bone marrow usually cause infection, dehydration and electrolyte imbalance. Death usually occurs within 2 weeks.

Cardiovascular (CV)/ Central Nervous System (CNS) syndrome- the full syndrome will usually occur with a dose greater than 50 Gy (5000 rads) though some symptoms may occur as low as 20 Gy or 2000 rads. Death occurs within 3 days. Death is likely due to collapse of the circulatory system as well as increased pressure in the confining cranial vault as the result of increased fluid content caused by edema, vasculitis and meningitis.

CHAPTER-12

ACUTE POISONING OF HYDROCARBONS

Hydrocarbon exposure may cause life threatening toxicity and in some cases sudden death.

Classification of hydrocarbon

Carbon and hydrogen atoms

- Aliphatic (open chain) and aromatic (benzene ring)

Household and occupational settings

- Fuels
- Lighter fluids
- Lamp oil
- Paints
- Paint removers
- Pesticides
- Medications
- Cleaning and polishing agents
- Spot removers
- Degreasers
- Lubricants
- Solvents

Most hydrocarbons result from petroleum distillation. Aliphatic mixtures of hydrocarbons of different chain lengths. Chain length and branching determines the phase of the hydrocarbon at room temperature. Short-chain (methane, propane or butane): gases. Intermediate-chain: liquids. Most hydrocarbon exposures seen in the ED. Long-chain: waxes/solids

- Wood distillates
    - Turpentine and pine oil
    - GI absorption greater than petroleum distillates
    - CNS depression
- Aromatics and halogenated aliphatic hydrocarbons
    - Industrial solvents
    - Inhalation route of toxicity
    - Substance abusers and some jobs most often affected
    - CNS, cardiovascular, hepatic, renal and hematologic toxicity
- Additives such as lead in gasoline and pesticides
    - Toxic additive usually dictates the clinical approach

Toxic potential of hydrocarbons depends on:
- Physical characteristics
    - Volatility, viscosity, surface tension
- Chemical characteristics
    - Aliphatic, aromatic, halogenated
- Presence of toxic additives
    - Pesticides, heavy metals
- Route of exposure
- Concentration & Dose

Aspiration Potential Depends On

Viscosity
- Lower viscosity, greater risk for aspiration
- Low
- Gasoline, kerosene, mineral seal oil, turpentine and aromatic and halogenated hydrocarbons
- High
- Diesel oil, grease, mineral oil, paraffin wax and petroleum jelly

Surface tension

    – Lower increases risk of aspiration

Volatility

    – Higher, increased risk of systemic absorption and toxicity

        • Aromatic hydrocarbons, halogenated hydrocarbons or gasoline

Acute toxicity of hydrocarbon affects the following organ system like Pulmonary, Neurologic, GIT, Cardiac, Hepatic, Renal, Hematologic, and Dermal.

Pulmonary Toxicity of Hydrocarbon

1° adverse affect of hydrocarbon exposure leads to cause pulmonary toxicity. Typically unintentional childhood ingestion (Small amounts of aliphatic hydrocarbons stored at home) is primary cause for toxicity. it shows limited GI absorption. Ingestion of aromatics or halogenated less likely to result in aspiration as GI absorption is greater. Risk and degree of aspiration not volume dependent. Occurs from aspiration into pulmonary tree like Occurs at time of ingestion and Hydrocarbons do not reflux into airway. Vomiting increases risk of aspiration. Pulmonary toxicity of hydrocarbon develop Pneumatoceles, Pneumothoraces, Pneumomediastinum, Bacterial superinfection, ARDS, Long-term pulmonary dysfunction and Death.

Pulmonary toxicity has been identified easily due to Irritation of oral mucosa and tracheobronchial tree cause following symptoms: Coughing, Choking, Gasping, Dyspnea, Burning of the mouth. Grunting respirations, Retractions, Tachypnea, Tachycardia, Cyanosis, Odor of hydrocarbons may be present.

CNS Toxicity

Systemic absorption of hydrocarbon through GIT, aspiration and dermal exposure leads to cause direct Central nervous system toxicity. CNS toxicity primarily cause hypoxia nature. The dose dependent effect of hydrocarbon is dizziness, slurred speech, ataxia, lethargy, obtundation, coma, apnea, exhilaration, giddiness, tremor, agitation, convulsions, confusion, hallucinations, psychosis and confused with alcohol intoxication.

Other CNS toxic effects are recurrent headaches, cerebellar ataxia, chronic encephalopathy, tremors, emotional lability, mental status changes, cognitive impairment, and psychomotor impairment. continuous exposure of hydrocarbon leads to cause peripheral polyneuropathy, demyelinization and retrograde axonal degeneration. The onset of symptoms may be delayed months to years. Long distal nerves most vulnerable and cause foot and wrist drop and numbness and paresthesias.

Gastrointestinal Toxicity

Most of the hydrocarbon act as intestinal irritants and cause burning in the mouth and throat, abdominal pain, belching, nausea, vomiting and diarrhea. In some cases corrosive GI injury and pancreatitis also reported.

Cardiac Toxicity

Halogenated and aromatic hydrocarbons produced ventricular tachycardia and ventricular fibrillation. Aliphatic hydrocarbon shows dysrhythmia and sudden death due to high sensitization of heart with catecholamines. Acute exposure cause decreased myocardial contractility, decreased peripheral vascular resistance, bradycardia, atrioventricular conduction blocks. Increased in mortality due to cardiac toxicity occur in association with asphyxia, respiratory depression, vagal inhibition.

Renal and Metabolic Toxicity

Acute exposure of hydrocarbon cause proteinuria, renal insufficiency, renal tubular acidosis, non-anion gap metabolic acidosis, hypokalemia, hypophosphatemia, rhabdomyolysis, high anion gap metabolic acidosis i.e., accumulation of hippuric and benzoic acid metabolites.

Hepatic Toxicity

Chronic exposure of carbon tetrachloride may result in cirrhosis and other hydrocarbon cause hepatic cell destruction via lipid peroxidation from free radicals, acute fatty degeneration centrilobular necrosis, liver function tests elevated 24 hours after ingestion with development of liver tenderness and jaundice in 48-96 hours.

Dermal Toxicity

Hydrocarbons are irritants and sensitizers to skin and cause pruritis, local erythema, papules, vesicles, generalized scarlatiniform eruption and exfoliative dermatitis.

Treatment

Pre-hospital Treatment

Not all ingestions require hospital evaluation. Less than 1% requires physician intervention. Asymptomatic after ingestion watched safely at home. Decision supported when: Ingestion is accidental, Known ingredients, Ingredients not significantly systemically toxic when ingested and reliable follow-up can be ensured. Symptomatic and intentional exposures should be referred to hospital for further evaluation. Accidental volatile exposure and abusers need cardiac monitoring and primary care transport due to potential of life-threatening dysrhythmias. In case of Hypotension: aggressive fluid resuscitation wants to given to the patient. Glucose, thiamine and naloxone should be considered in cases of altered mental status.

If any patient cloths contaminated with hydrocarbon fully undress the patient to prevent ongoing contamination from hydrocarbon-soaked clothes. The handling staffs want to wear gloves, goggles and aprons prevent possible secondary exposure. Affected skin washed with soap and running water. Affected eye irrigated with normal saline.

GIT decontamination

Need depends on type of hydrocarbon and route of exposure. Little data as to effectiveness of GI decontamination. For most ingestions GI decontamination of little benefit. Most aliphatic HC ingestions do not require GI decontamination due to poor GI absorption. Supportive care and treatment for coexisting ingestions also required. Risk vs. benefits of GI decontamination is systemic toxicity by intestinal absorption and risks of aspiration associated with gastric emptying. Suicidal ingestions involve large amounts of HCs and associated with spontaneous emesis. So, further decontamination not usually required. Ipecac induced emesis contraindicated for patients and should be avoided. Charcoal not recommended for most hydrocarbon ingestions. Because it distends the stomach increasing the risk for vomiting and aspiration. Cathartics no proven efficacy in hydrocarbons because many hydrocarbon ingestion already have diarrhea. Oil based cathartics contraindicated and cause increased GI absorption with risk of lipoid pneumonia when aspirated.

Pulmonary Treatment

Nebulized oxygen helpful to maintain the breathing during pre hospital management. Inhaled β_2 agonists is useful for bronchospasm. Administration of steroids contraindicated during respiratory depression and may cause impairment of cellular immune response and Increased chance of bacterial superinfection. Antibiotics are beneficial for superimposed bacterial pneumonitis.

CHAPTER-13

CAUSTICS POISON

Classification

Acids

- Cleaners: HCl, H_2SO_4
- Etching and metal cleaning: HF
- Metal Plating: Chromic acid
- Leather and Textile tanning: Formic acid

Alkali

- Cleaning fluids: NaOH, KOH
- Concrete: CaOH
- Photography: LiOH
- Fertilizer: Ammonium hydroxide

Sources

Caustics are commonly available as a source from household items. Caustics are less concentrate than industry. Acids like sulfuric acid obtained from drain cleaners and automobile batteries, hydrochloric acid present in cleaners, airplane glue contain formic acid and hydrogen fluoride present in rust removers. Alkali like NaOH present in drain cleaners, oven cleaners, clinitest tablets, sodium hypochlorite present in household bleach which is most common alkali to cause benign condition for the exposures, and ammonium present in glass, tub and tile cleaners.

Acute poisoning effect of alkali

Acute exposure of alkali may cause deep liquefaction necrosis. The stages of necrosis like proteins rapidly denatured, lipids undergo saponification, cellular destruction on contact, thrombosis of microvasculature which leads to further necrosis. The initial solid alkali exposure cause esophageal injury. Severe intentional ingestion may result in multisystem organ injury like gastric perforation, necrosis of abdominal viscera and pancreas, gallbladder and small intestine injury.

Household bleach contains 3-6% sodium hypochlorite solution (pH of 11) which is not corrosive to esophagus and ingestion may cause emesis with 2° to gastric or pulmonary irritation. Industrial bleach contains higher concentrations of sodium hypochlorite which may cause esophageal necrosis with ingestion and aspiration pneumonitis.

Acute poisoning effect of acid

Strong acids produce coagulation necrosis, tissue destruction and cell death results in eschar formation which protects against deeper injury. Acid ingestion does not cause esophageal sparing and may settle in stomach which leads to cause gastric necrosis, perforation and hemorrhage. Acid poisoning is less tissue destruction than alkali and higher mortality than alkali ingestion may be due to complications of systemic absorption like metabolic acidosis, hemolysis and liver failure.

Clinical Features of acid and alkali poisoning

Severe pain, odynophagia, dysphonia, oral and facial burns, respiratory distress, abdominal pain, drooling, coughing and vomiting. Laryngotracheal injury causes dysphonia, stridor and respiratory distress. Esophageal and GI injury cause dysphagia, odynophagia, epigastric pain and vomiting.

Management

Initial management

Initial step is airway evaluation may have oral, pharyngeal or larygnotracheal injury. Ideally should have fiberoptic evaluation prior to intubation to determine extent of damage. Blind nasotracheal intubation is contraindicated due to risk of further injury. Establish airway early to avoids 2° effects of injury such as edema. Oral intubation with direct visualization is the first choice for definitive management. Surgical cricothyrotomy may be required

Gastric Decontamination

Charcoal does not bind caustics well. Ipecac is contraindicated and may cause vomiting may precipitates perforation and results in repeated exposure of airway and GI tract to caustic agent. Application of nasogastric tube is high risk of perforation with alkali ingestion. Endoscopist may insert with acid ingestion to aspirate residual. Endoscopy must be done to locate the severity of injury after post ingestion. Endoscopy within first several hours after ingestion may be beneficial to start the treatment. Computor Tomography or Ultra Sound may be used and may screen for intraabdominal necrosis outside the GI tract or in areas not reachable with endoscopy.

CHAPTER-14

MUSHROOM POISONING

Introduction

Wild mushrooms that grow in forests and meadows, are of various types, and it is common for the local population to consume them. Nevertheless, mushrooms are one of the most common toxic exposures, with over 12000 mushroom exposures reported to poison centers in 1996, or roughly 5 for every 100000 population. While most mushroom ingestions do not cause a clinically significant toxidrome, the lethal potentials of a select few make mushroom toxicity an important subject. Ingestion is the most common route of entry, but intravenous injections of mushroom toxins and inhalations of mushroom spores have been reported.

The symptoms and signs of mushroom poisoning range from mild gastrointestinal symptoms to organ failure and death. Toxicity may also vary based on the amount ingested, the age of the mushroom, the season, the geographic location, and the way in which the mushroom has been prepared prior to ingestion. Eating poisonous mushrooms can cause various types of reactions, such as allergic gastroenteritis, psychological relaxation and fatal liver intoxication. The pathogenicity of these mushrooms depends on the cyclopeptide toxins. Despite these measures, the species is unknown in > 90% of digestion. However, Amantia species are responsible for the vast majority of deaths.

Mushroom toxidromes are classified according to toxins and clinical presentations. Mushroom toxins have been divided into the following 7 main categories:

- Amatoxins (cyclopeptides)
- Orellanus (Cortinarius species)
- Gyromitrin(monomethylhydrazine)

- Muscarine
- Ibotenic acid
- Psilocybin
- Coprine(disulfiramlike)

Mushrooms of the genera Amanita and Lepiota contain amatoxins, which are thermostable and bicyclic peptide toxins. Amanitin phalloides syndrome or Mycetismus choleriformis accounts for 90-95% of all fatalities from mushroom poisoning in North America. This discussion follows a clinical format because the offending mushroom is frequently unavailable for identification and poisoning may occur from a single species or a combination of different species. Trestrail's data indicate that the mushroom was available for identification in only 3.4% of exposures. Amatoxins, especially amanitin, are absorbed by the gut and degrade the cells of liver and kidneys.

Management

There are nine general groupings useful for clinical management. These groups of toxin can be divided into early toxicity (within 1 hour after ingestion) and delayed toxicity (6 hour to 20 days). The groups causing early onset of symptoms include Coprine, GI toxin, Ibotenic acid, muscimol, Muscarine and Psilocybin. The other groups causing late onset of symptoms include Cyclopeptides, Orellanus and Gyromitrin.

Treatment

Asymptomatic patients

1. When suspecting mushroom poisoning, try to get the specimen as possible as you can, and then contact a regional poison control center.

2. Give the patients 0.5-1.0 g/kg of activated charcoal orally and intravenous fluid to prevent dehydration or electrolytes imbalance

3. Monitor patients in the emergency department for more than four hours. If patients remain asymptomatic, you can discharge them with adequate instructions. Advise patients to contact the hospital immediately if they

become symptomatic or have any discomfort. If the mushroom is identified as potentially toxic or the patient becomes symptomatic, admission to the hospital is recommended.

Symptomatic patients

1. The basic elements of supportive care are critical in the evaluation and management of the poisoned patients. The priority must be as following : airway, breathing, circulation.

2. Obtain the mushroom specimen as possible as you can.

3. Consider gastric lavage with activated charcoal every 2-6 hours.

4. Cardiopulmonary monitoring should be available.

5. Closely monitor fluid, electrolytes, and glucose status and correct them. Rehydrate with isotonic fluids. Forced dieresis is not recommended.

6. If amanitin ingestion is suspected or proven, careful attention to clotting studies and renal and hepatic profiles is important. Early consultation with a medical toxicologist is recommended.

7. Intensive unit care may be necessary when the patient's condition is poor.

Special consideration

Amatoxin

Amatoxins within the mushrooms are the most common fatal conditions in mushroom poisonings. It has a latent period of 6-12 hours after digestion. At the period, the patient may have GI symptoms. Hepatic and renal failure may be encountered. Deaths may occur in 3-7 days. Mortality rates range from 10-60%. Some drugs may be useful according to animal experiments, but only anecdotal support is available for humans.

1. High doses of penicillin (300,000 to 1,000, 000 U/kg/day) are required to decrease toxicity.

2. Vitamin K (if coagulopathy is present)

3. Silybinin (water-soluble milk thistle extract, not available in the US)

4. Hyperbaric oxygen

5. High-dose cimetidine, vitamin C, zinc, and thiol compounds are useful in animal models.

Neurological symptoms

Psilocybin or psilocin toxins are neuroactive chemicals similar to lysergic acid diethylamide (LSD).

Anti-cholinergic symptoms

The symptoms include tachycardia, hypertension, warm, dry skin and mucous membranes, and mydriasis. When patients have anti-cholinergic symptoms, physostigmine may be considered.

Muscarinic symptoms

Muscarinic symptoms are characterized by the "SLUDGE" syndrome. The "SLUDGE" are as following: salivation, lacrimation, urination, defecation, GI hypermotility, and emesis. Atropine can be used for bradycardia and hypotension. Oxygen and inhaled beta-agonists are also helpful in treating patients with increased pulmonary secretions and bronchospasm.

Renal failure

Nephrotoxins in mushrooms are norleucine and chlorocrotylgycine. Patients digesting nephrotoxin-contained are usually asympots. Supportive hemodialysis may be required in 30-50% patients. The others recover without sequalae.

Accompanying alcohol digestion

A kind of mushroom, Coprinus genus, contains coprine which is chemically related to disulfiram. The toxin inhibits alcohol dehydrogenase 2 hours after digestion, and the effect may last up to 72 hours. When the patients digest alcohol with this kind of mushroom, the major symptoms are facial flushing, headache, tachycardia, nausea and vomiting.

Conclusion

The species of mushroom are numerous. There are various clinical presentations depending on the ingested species. Most ingested species remain unknown.

Treatments include gastric decontamination (activated charcoal), observation of 12 to 24 hours for delayed-onset symptoms (which may indicate serious toxicity), laboratory studies, intravenous fluid hydration and supportive care. In general, most cases of the mushroom poisonings are mild to moderate gastrointestinal upset. There are no rules available about treating mushroom intoxication or identifying mushroom toxin in the emergency department, so the diagnosis and treatment must be based on the history of ingestion and associated clinical presentations.

CHAPTER-15

MYCOTOXIN

Secondary metabolites (chemicals) of a fungus that produce toxic results in another organism. Cytotoxic: disrupt cell structures such as membranes, and processes such as protein, DNA, and RNA synthesis. Lack of visible appearance of fungus does not negate presence of mycotoxins. Toxins can remain in the organism after fungus has been removed. Less selective in organism selection, can cross plant species barrier. Can be heat stable, not destroyed by canning or other processes.

History of mycotoxin

Mycotoxin contamination has affected humans for thousands of years. In 7th and 8th century, festival for Roman God Robigus, protector of grain and trees was celebrated to stave off rust and mold. Middle Ages had outbreaks of ergotism. Only in last 30-40 years have scientists been able to isolate specific toxins from their fungal source. Research ideas and methodologies, in this field, change frequently, and data from 20 years ago are considered questionable.

Symptoms of mycotoxicosis

1. Drugs and antibiotics are not effective in treatment.
2. The symptoms can be traced to foodstuffs or feed.
3. Testing of said foodstuffs or feed reveals fungal contamination.
4. The symptoms are not transmissable person to person.
5. The degree of toxicity is subject to person's age (more often in very young and very old), sex (more often in females than males) and nutritional status.
6. Outbreaks of symptoms appear seasonally.
7. Economic loss due to impaired health of stock animals.

8. Illness: symptoms can include cold/flu-like symptoms, sore throats, headaches, nose bleeds, fatigue, diarrhea, dermatitis, and immune suppression, and vary by species.

9. Death.

Origin of principal mycotoxin occurring in common feeds and forages

Mycotoxins	Fungal species
Aflatoxins	Aspergilus flavus; A. Parasiticus
Cyclopiazonic acid	A. flavus
Ochratoxin A	A. ochraceus; Penicillium viridicatum; P. cyclopium
Citrinin	P. citrinum; P. expansum
Patulin	P. expansum
Citreovirdin	P. citreo virde
Deoxynivalenol	Fusarium culmorum; F. graminearum
T-2 toxin	F. sperotrichioides; F. poae
diacetoxyscirpenol	F. sperotrichioides; F. Graminearum; F. poae
Zearalenone	Fusarium culmorum; F. graminearum

Aflatoxins

This group includes aflatoxin B1, B2, G1 and G2 (AFB1, AFB2, AFG1 and AFG2, respectively). In addition, aflatoxin M1 (AFM1) has been identified in the milk of dairy cows consuming AFB1-contaminated feeds. The aflatoxigenic *Aspergilli* are generally regarded as storage fungi, proliferating under conditions of relatively high moisture/humidity and temperature. Aflatoxin contamination is, therefore, almost exclusively confined to tropical feeds such as oilseed by-products derived from groundnuts, cottonseed and palm kernel. Aflatoxin contamination of maize is also an important problem in warm humid regions where *A. flavus* may infect the crop prior to harvest and remain viable during storage.

Surveillance of animal feeds for aflatoxins is an ongoing issue, owing to their diverse forms of toxicity and also because of legislation in developed countries (D'Mello and Macdonald, 1998). In the United Kingdom, analysis conducted during the 1987-1990 period indicated that all imported feedstuffs complied with legislation in force for

AFB1 levels. Elsewhere, however, aflatoxin levels in certain feeds still pose serious risks to animal health. In China, 85 percent of maize samples were contaminated with both AFB1 and fumonisin B1 at levels ranging from 8 to 68 g/kg and 160 to 25970 g/kg, respectively. Feed-grade maize in northern Viet Nam had AFB1 levels ranging from 9 to 96 g/kg, and fumonisin B1 levels in the range of 271 to 3 447 g/kg (Placinta, D'Mello and Macdonald, 1999). Between 1988 and 1989, analyses of farmgate milk in the United Kingdom showed low levels of AFM1 contamination, but more than 50 percent of milk samples in the United Republic of Tanzania were found to contain the mycotoxin (D'Mello and Macdonald, 1998). The importance of aflatoxins in anima health emerged in 1960, following an incident in the United Kingdom in which 100 000 turkey poults died from acute necrosis of the liver and hyperplasia of the bile duct ("turkey X disease"), attributed to the consumption of groundnuts infected with *Aspergillus flavus*. This event marked a defining point in the history of mycotoxicoses, leading to the discovery of the aflatoxins. Subsequent studies showed that aflatoxins are acutely toxic to ducklings, but ruminants are more resistant. However, the major impetus arose from epidemiological evidence linking chronic aflatoxin exposure with the incidence of cancer in humans.

Ochratoxins

The *Aspergillus* genus includes a species (*A. ochraceus*) that produces ochratoxins, a property it shares with at least two *Penicillium* species. Ochratoxin A (OA) and ochratoxin B are two forms that occur naturally as contaminants, with OA being more ubiquitous, occurring predominantly in cereal grains and in the tissues of animals reared on contaminated feed. Another mycotoxin, citrinin, often co-occurs with ochratoxin. In recent Bulgarian wheat samples, OA and citrinin levels ranged from < 0.5 to 39 g/kg and from < 5 to 420 g/kg, respectively. In oats, higher levels of OA were detected (maximizing at 140 g/kg) while citrinin was below detection limits (D'Mello, 2001).

The ochratoxins and citrinin are nephrotoxic to a wide range of animal species. OA is frequently implicated in porcine nephropathy and in Balkan endemic nephropathy of humans. The role of citrinin in these syndromes has yet to be elucidated.

Fusarium mycotoxins

Extensive data now exist to indicate the global scale of contamination of cereal grains and animal feed with *Fusarium* mycotoxins (D'Mello and Macdonald, 1998). Of particular importance are the trichothecenes, zearalenone (ZEN) and the fumonisins. The trichothecenes are subdivided into four basic groups, with types A and B being the most important. Type A trichothecenes include T-2 toxin, HT-2 toxin, neosolaniol and diacetoxyscirpenol (DAS). Type B trichothecenes include deoxynivalenol (DON, also known as vomitoxin), nivalenol and fusarenon-X. The production of the two types of trichothecenes is characteristic for a particular *Fusarium* species. However, a common feature of the secondary metabolism of these fungi is their ability to synthesize ZEN which, consequently, occurs as a co-contaminant with certain trichothecenes. The fumonisins are synthesized by another distinct group of *Fusarium* species (Table 1).

Three members of this group (fumonisins B1, B2 and B3) often occur together in maize. Virtually all the toxigenic species of *Fusarium* listed in Table 1 are also major pathogens of cereal plants, causing diseases such as head blight in wheat and barley and ear rot in maize. Harvested grain from diseased crops is therefore likely to be contaminated with the appropriate mycotoxins, and this is supported by ample evidence. Surveillance of grain and animal feed for the occurrence of *Fusarium* mycotoxins has been the subject of many investigations over recent years (Tables 2 and 3). The global distribution of these mycotoxins is a salient feature, but striking regional differences should also be noted. Another aspect worthy of comment is consistent evidence of the co-occurrence of various *Fusarium* mycotoxins in the same sample. These issues have been considered at greater length by Placinta, D'Mello and Macdonald (1999) who, for example, referred to a German study in which 94 percent of wheat samples analysed were contaminated by between two and six *Fusarium* mycotoxins and 20 percent of the samples were co-contaminated with DON and ZEN (Table 2). The most frequent combination included DON, 3-ADON and ZEN. T-2 and HT-2 toxins were detected at levels ranging from 0.003 to 0.250 mg/kg and 0.003 to 0.020 mg/kg, respectively, but these mycotoxins only occurred in combination with DON, NIV and ZEN.

Endophyte alkaloids

The endophytic fungus *Neotyphodium coenophialum* occurs in close association with perennial tall fescue, while another related fungus, *N. lolii*, may be present in perennial ryegrass (D'Mello, 2000). Ergopeptine alkaloids, mainly ergovaline, occur in *N. coenophialum*-infected tall fescue, while the indole isoprenoid lolitrem alkaloids, particularly lolitrem B, are found in *N. lolii*-infected perennial ryegrass. The ergopeptine alkaloids reduce growth, reproductive performance and milk production in cattle, while the lolitrem compounds induce neurological effects in ruminants.

Phomopsins

In Australia, lupin stubble is valued as fodder for sheep, but infection with the fungus *Phomopsis leptostromiformis* is a major limiting factor because of toxicity arising from the production of phomopsins by the fungus. Mature or senescing parts of the plant, including stems, pods and seeds, are particularly prone to infection. Phomopsin A is considered to be the primary toxin, causing effects such as ill-thrift, liver damage, photosensitization and reduced reproductive performance in sheep (D'Mello and Macdonald, 1998).

Sporidesmin

Pithomyces chartarum is a ubiquitous saprophyte of pastures and has the capacity to synthesize sporidesmin A, a compound causing facial eczema and liver damage in sheep.

CHAPTER-16

FOOD POISONING

Food poisoning syndrome results from ingestion of water and wide variety of food contaminated with pathogenic microorganisms (bacteria, viruses, protozoa, fungi), their toxins and chemicals. Food poisoning must be suspected when an acute illness with gastrointestinal or neurological manifestation affect two or more persons, who have shared a meal during the previous 72 hours. The term as generally used encompasses both food-related infection and food-related intoxication.

Some microbiologists consider microbial food poisoning to be different from food-borne infections. In microbial food poisoning, the microbes multiply readily in the food prior to consumption, whereas in food-borne infection, food is merely the vector for microbes that do not grow on their transient substrate. Others consider food poisoning as intoxication of food by chemicals or toxins from bacteria or fungi. Consumption of poisonous mushroom leads to mycetism, while consumption of food contaminated with toxin producing fungi leads to mycotoxicosis.

Some microorganisms can use our food as a source of nutrients for their own growth. By growing in the food, metabolizing them and producing by-products, they not only render the food inedible but also pose health problems upon consumption. Many of our foods will support the growth of pathogenic microorganisms or at least serve as a vector for their transmission. Food can get contaminated from plant surfaces, animals, water, sewage, air, soil, or from food handlers during handling and processing.

Classification of Food Poisoning:

Based on symptoms and duration of onset

1. Nausea and vomiting within six hours (*Staphylococcus aureus*, *Bacillus cereus*)
2. Abdominal cramps and diarrhoea within 8-16 hours (*Clostridium perfringens, Bacillus cereus*)
3. Fever, abdominal cramps and diarrhoea within 16-48 hours (Salmonella, Shigella, *Vibrio parahemolyticus*, Enteroinvasive E.coli, Campylobacter jejuni)
4. Abdominal cramps and watery diarrhoea within 16-72 hours (Enterotoxigenic *E.coli*, *Vibrio cholera* O1, O139, *Vibrio parahemolyticus*, NAG vibrios, Norwalk virus)
5. Fever and abdominal cramps within 16-48 hours (*Yersinia enterocolitica*)
6. Bloody diarrhoea without fever within 72-120 hours (Enterohemorrhagic *E.coli* O157:H7)
7. Nausea, vomiting, diarrhoea and paralysis within 18-36 hours (*Clostridium botulinum*)

Based on pathogenesis

1. Food intoxications resulting from the ingestion of preformed bacterial toxins. (*Staphylococcus aureus, Bacillus cereus, Clostridium botulinum, Clostridium perfringens*)
2. Food intoxications caused by noninvasive bacteria that secrete toxins while adhering to the intestinal wall (Enterotoxigenic *E.coli, Vibrio cholerae, Campylobacter jejuni*)
3. Food intoxications that follow an intracellular invasion of the intestinal epithelial cells. (Shigella, Salmonella)
4. Diseases caused by bacteria that enter the blood stream via the intestinal tract. (*Salmonella typhi, Listeria monocytogenes*).

Bacterial Etiology Of Food Poisoning:

Food infections by bacteria can be divided into two types:

1. Those in which the food does not ordinarily support the growth of pathogens but merely carries them. E.g. Salmonella, Shigella, Vibrio etc.
2. those in which the food can serve as a culture medium for growth of pathogens to numbers that can infect the person.

Food borne infections by bacteria can also be classified as toxicosis and food-infections. In toxicosis, the toxins are released by bacteria such as Clostridia, Bacillus and Staphylococcus. In food-infections, the bacteria are ingested, which later initiate the infection.

Staphylococcus aureus

S.aureus is gram positive cocci that occurs in singles, pairs, short chains, tetrads and irregular grape like clusters. It is present ubiquitously in the environment. Only those strains that produce enterotoxin can cause food poisoning. Food is usually contaminated from infected food handler. The food handler with an active lesion or carriage can contaminate food.

Incriminated food:

Custard and cream filled bakery food, ham, chicken, meat, milk, fish, salads, puddings, pie etc.

Pathogenesis:

If the food is stored for some time at room temperature the organism may multiply in the food and produce toxin. The bacteria produce enterotoxin while multiplying in food. *S.aureus* is known to produce six serologically different types of enterotoxins (A, B, C, C2, D and E) that differ in toxicity. Most food poisoning is caused by enterotoxin A. Isolates commonly belong to phage type III. These enterotoxins tend to be heat stable, with type B being most heat resistant. Low temperature heat inactivated enterotoxin can undergo reactivation in some food. Ingestion of as little as 23 µg of enterotoxin can induce vomiting and diarrhoea. Staphylococcal enterotoxins act as superantigens, binding to MHC II molecules and stimulating T cells to divide and produce lymphokines such as IL-2 and TNF-alpha, which induces diarrhea. The toxin acts on the receptors in the gut and sensory stimulus is carried to the vomiting center in the brain by vagus and sympathetic nerves.

Incubation period:

Since the ingested food contains preformed toxin, the incubation period is usually 1-6 hours.

Clinical features:

The onset is sudden and is characterized by vomiting and diarrhea but no fever. The illness lasts less than 12 hours. There are no complications and treatment is usually not necessary.

Laboratory diagnosis:

The presence of a large number of *S.aureus* organisms in a food may indicate poor handling or sanitation; however, it is not sufficient evidence to incriminate a food as the cause of food poisoning.

Staphylococcal food poisoning can be diagnosed if they are isolated in large numbers from the food and their toxins demonstrated in the food or the isolated *S.aureus* must be shown to produce enterotoxins. Dilutions of food may be plated on Baird-Parker agar or Mannitol Salt agar. Enterotoxin may be detected and identified by gel diffusion.

Bacillus cereus:

B.cereus is a gram positive aerobic spore bearing bacilli. It is found abundantly in environment and vegetation.

Incriminated food:

Commonly associated with rice and vegetables.

Pathogenesis:

During the slow cooling, spores germinate and vegetative bacteria multiply, then they sporulate again. Sporulation is also associated with toxin production. The toxin is heat-stable, and can easily withstand the brief high temperatures used to cook fried rice. The short-incubation form is most often associated with fried rice that has been cooked and then held at warm temperatures for several hours. Long-incubation food poisoning is frequently associated with meat or vegetable-containing foods after cooking. The short-incubation form is caused by a preformed heat-stable enterotoxin of molecular weight less than 5,000 daltons. The long- incubation form of illness is mediated by a heat-labile enterotoxin (molecular weight of approximately 50,000 daltons), which activates intestinal adenylate cyclase and causes intestinal fluid secretion.

Incubation period:

1-6 hours in short-incubation form and 8-16 hours in long incubation form.

Clinical features:

B. cereus causes two types of food-borne intoxications. The 'emetic-type' or the short incubation type has an incubation period of 1 to 6 hours. It is characterized by nausea, vomiting and abdominal cramps and resembles *S. aureus* food poisoning in its symptoms and incubation period. Within 16 hours of eating contaminated fried rice, patients suffer a bout of vomiting that generally lasts for less than a day. The second type is manifested primarily by abdominal cramps and diarrhea with an incubation period of 8 to 16 hours. Diarrhea may be a small volume or profuse and watery. This type is referred to as the "long-incubation" or diarrheal form of the disease, and it resembles food poisoning caused by *Clostridium perfringens*. In either type, the illness usually lasts less than 24 hours after onset.

Laboratory diagnosis:

The short-incubation or emetic form of the disease is diagnosed by the isolation of B. cereus from the incriminated food. The long-incubation or diarrheal form is diagnosed by isolation of the organism from stool and food. Isolation from stools alone is not sufficient because 14% of healthy adults have been reported to have transient gastrointestinal colonization with B. cereus.

Clostridium perfringens:

It is a gram positive anaerobic spore bearing bacilli that is present abundantly in the environment, vegetation, sewage and animal feces.

Incriminated food: food-borne outbreaks of *C.perfringens* involve meat products that are eaten 1- 2 days after preparation. Meats that have been cooked, allowed to cool slowly, and then held for some time before eating are commonly incriminated. Fish pastes and cold chicken too have been incriminated.

Pathogenesis: Spores in food may survive cooking and then germinate when they are improperly stored. When these vegetative cells form endospores in the intestine, they release enterotoxins. The bacterium is known to produce at least 12 different toxins. Food poisoning is mainly caused by Type A strains, which produces alpha and theta toxins. The toxins result in excessive fluid accumulation in the intestinal lumen.

Incubation period: 8-24 hours

Clinical features: Illness is characterized by acute abdominal pain, diarrhea, and vomiting. Illness is self- limiting and patient recovers in 18-24 hours.

Laboratory diagnosis: Since the bacterium is present normally in the intestine, their isolation from feces may not be sufficient to implicate it. Similarly, isolation from food except in large numbers (>105/gram) may not be significant. The homogenized food is diluted and plated on selective medium as well as Robertson cooked meat medium and incubated anaerobically. The isolated bacteria must be shown to produce enterotoxin.

Clostridium botulinum:

It is a gram positive anaerobic spore bearing bacilli that is widely distributed in soil, sediments of lakes and ponds, and decaying vegetation.

Incriminated food: Most cases of botulism are associated with home canned or bottled meat, vegetables and fish. In general, the low and medium acid canned foods are often incriminated. The anaerobic environment produced by the canning process may further encourage the outgrowth of spores.

Pathogenesis: Not all strains of *C.botulinum* produce the botulinum toxin. Seven toxigenic types of the organism exist, each producing an immunologically distinct form of botulinum toxin. The toxins are designated A, B, C1, D, E, F, and G. Lysogenic phages encode toxin C and D serotypes. Food-borne botulism is not an infection but an intoxication since it results from the ingestion of foods that contain the preformed clostridial toxin. If contaminated food has been insufficiently sterilized or canned improperly, the spores may germinate and produce botulinum toxin. The toxin is released only after the death and lysis of cells. The toxin resists digestion and is absorbed by the upper part of the GI tract and then into the blood. It then reaches the peripheral neuromuscular synapses where the toxin binds to the presynaptic stimulatory terminals and blocks the release of the neurotransmitter acetylcholine. This results in flaccid paralysis. Even 1-2 µg of toxin can be lethal to humans.

Incubation period: 12-36 hours

Clinical features: Common features include vomiting, thirst, dryness of mouth, constipation, ocular paresis (blurred-vision), difficulty in speaking, breathing and swallowing. Coma or delirium may occur in some cases. Death may occur due to respiratory paralysis within 7 days.

Laboratory diagnosis: Spoilage of food or swelling of cans or presence of bubbles inside the can indicate clostridial growth. Food is homogenized in broth and inoculated in Robertson cooked meat medium and blood agar or egg-yolk agar, which is incubated anareobically for 3-5 days at 37oC. The toxin can be demonstrated by injecting intraperitoneally the extract of food or culture into mice or guinea pig.

Enterotoxigenic *E.coli* (ETEC):

E.coli are gram negative enteric bacilli that are carried normally in the intestine of humans and animals. Some specific serotypes harbor plasmids that code for toxin production. The enterotoxin production is limited to following O serotypes: O6, O8, O15, O25, O63, O78, O148 and O159.

Incriminated food: Infection is acquired by ingestion of food or water contaminated with ETEC. Contamination of water with human sewage may lead to contamination of foods. Infected food handlers may also contaminate foods. The infective dose is 10^6-10^{10} bacilli.

Pathogenesis: The bacteria colonize the GI tract by means of fimbriae to specific receptors on enterocytes of the proximal small intestine. Enterotoxins produced by

ETEC include the LT (heat-labile) toxin and or the ST (heat-stable) toxin. LTs are similar to cholera toxin in structure and mode of action. LTs are holotoxin consisting of A subunit and B subunit. The B subunit of LTs binds to specific ganglioside receptors (GM1) on the epithelial cells of small intestine and facilitates the entry of A subunit where it activates adenylate cyclase. Stimulation of adenylate cyclase causes an increased production of cAMP, which leads to hypersecretion of water and electrolytes into the lumen and inhibition of sodium reasborption.

Incubation period: 16-72 hours

Clinical features: Sudden onset of watery diarrhea associated with nausea, vomiting, abdominal cramping and bloating is commonly observed. This bacterium is responsible for majority of traveler's diarrhea. The disease is self-limiting and resolves in few days.

Laboratory diagnosis: The sample of feces is cultured on McConkey's agar. The ETEC stains are indistinguishable from the resident *E.coli* by biochemical tests. These strains are differentiated from nontoxigenic *E.coli* present in the bowel by a variety of in vitro immunochemical, tissue culture, or DNA hybridization tests designed to detect either the toxins or genes that encode for these toxins. With the availability of a gene probe method, foods can be analyzed directly for the presence of enterotoxigenic E.coli in about 3 days. LTs can be detected by Ligated rabbit ileal loop test, morphological changes in Chinese hamster overy cells and Y1 adrenal cells, ELISA, immunodiffusion, coaglutination etc.

Enterohemorrhagic *E.coli* (EHEC):

E.coli are gram negative enteric bacilli that are carried normally in the intestine of humans and animals. EHEC strains have been associated with many serogroups including O4, O26, O45, O91, O111, O145 and O157. The most serotype is O157:H7.

Incriminated food: Cattle appear to be the main source of infection; most cases being associated with the consumption of undercooked beefburgers and similar foods.

This disease is often associated with ingestion of inadequately cooked hamburger meat, raw milk, cream and cheeses made from raw milk.

Pathogenesis: EHEC strains may produce one or more types of cytotoxins, which are collectively referred as Shiga-like toxins (SLT) since they are antigenically and functionally similar to Shiga toxin produced by *Shigella dysenteriae*. SLTs were previously known as verotoxin. The toxins provoke cell secretion and kill colonic epithelial cells.

Incubation period: 72-120 hours

Clinical features: Initial symptoms may be diarrhea with abdominal cramps, which may turn into grossly bloody diarrhoea in a few days. There is however, no fever.

Laboratory diagnosis: Laboratory diagnoses involve culturing the faeces on McConkey's agar or on sorbitol McConkey's agar, where they don't ferment sorbitol. Strains can then be identified by serotyping using specific antisera. SLTs can be detected by ELISA and genes coding for them can be detected by DNA hybridization techniques.

Vibrio parahemolyticus:

They are straight or curved gram negative halophilic bacilli. In morphology and staining it resembles *V.cholerae* and is actively motile in liquid cultures. It is commonly found in coastal seas, where it has been isolated from marine fauna such as crabs, shrimps, fishes and molluscs.

Incriminated food: Infections are associated with consumption of uncooked or undercooked crabs, prawns, shrimps and other seafoods.

Pathogenesis: No enterotoxin has been demonstrated in the bacterium. The infection is thought to result from invasion of intestinal epithelium.

Incubation period: 7-48 hours

Clinical features: The clinical infection is characterized by a sudden onset of acute gastroenteritis. Infection may also result in diarrhoea, abdominal pain, vomiting and fever.

Laboratory diagnosis: Homogenized food may be inoculated into TCBS agar or into double strength alkaline peptone water and incubated overnight at 37oC. This bacterium is positive for Kanagawa phenomenon where isolates from human feces show hemolysis on blood agar.

Salmonella enteritidis:

These are gram negative rod shaped bacteria that are classified under family enterobacteriaceae. This species does not occur normally in humans but several animals act as reservoirs.

Incriminated food: Most important sources are chicken and poultry. Chicken, duck, turkey and goose may be infected with Salmonella, which then find its way into its feces, eggs or flesh of dressed fowl. Milk and milk products including ice creams have been incriminated.

Pathogenesis: Organism penetrates and passes through the epithelial cells lining the terminal portion of the small intestine. Multiplication of bacteria in the lamina propria produces inflammatory mediators, recruits neutrophils and triggers inflammation. Release of LPS causes fever. Inflammation causes release of prostaglandins from epithelial cells. Prostaglandins cause electrolytes to flow into lumen of the intestine. Water flows into lumen in response to osmotic imbalance resulting in diarrhea.

Incubation period: 12-36 hours

Clinical features: Sudden onset of abdominal pain, nausea, vomiting and diarrhea, which may be watery, greenish and foul smelling. This may be preceded by headache and chills. Other findings include prostration, muscular weakness and moderate fever. In most cases the symptoms resolve in 2-3 days without any complications.

Laboratory diagnosis: Homogenized food is cultured in selenite F broth and then sub-cultured on deoxycholate citrate agar. Plates are incubated at 37oC overnight and growth identified by biochemical tests and slide agglutination test.

Yersinia enterocolitica:

It is a gram negative psychrophilic rod shaped bacterium that is motile only at temperature below 30oC. *Yersinia enterocolitica* is widely distributed in environment and have been isolated frequently from soil, water and animals. The major animal reservoir for *Y.enterocolitica* strains that cause human illness is pigs, but may also found in many other animals including rodents, rabbits, sheep, cattle, horses, dogs, and cats. Serogroups that predominate in human illness are O:3, O:8, O:9, and O:5.

Incriminated food: Infection is most often acquired by eating contaminated food, especially raw or undercooked pork products. Drinking contaminated unpasteurized milk or untreated water can also transmit the infection.

Pathogenesis: This organism may survive and grow during refrigerated storage. Strains that cause human yersiniosis carry a plasmid that is associated with a number of virulence traits. Ingested bacteria adhere and invade M cells or epithelial cells. They exhibit resistance to complement and phagocytosis. They produce ST only at temperatures below 30°C. The role of ST in the disease process remains uncertain.

Incubation period: 4-7 days

Clinical features: Disease produced by *Y.enterocolitica* is a typical gastroenteritis characterized by fever, abdominal pain, and diarrhea, which is often bloody. Illness generally lasts from 1 to 2 weeks but chronic cases may persist for up to a year. Apart from gastroenteritis it may also cause pseudoappendicitis, mesenteric lymphadenitis, and terminal ileitis.

Laboratory diagnosis: Suspected food is homogenized in phosphate-buffered saline and inoculated into selenite F broth and held at 4oC for six weeks. The broth is sub-cultured at weekly intervals on DCA or Yersinia selective agar plates. This is termed as cold enrichment technique.

Campylobacter jejuni:

These are small, curved-spiral gram negative bacilli with polar flagella. *Campylobacter jejuni* appear in comma, S-shaped or "gull-wings/sea-gull" form. Campylobacter are harbored in reproductive and alimentary tracts of some animals.

Incriminated food: Transmission to humans occurs via a fecal-oral route, originating from farm animals, birds, dogs, and processed poultry, with chicken preparation comprising 50-70% of all campylobacter infections. The organism is transmitted to man in milk, meat products and contaminated water. Undercooked poultry and unpasteurized dairy are most often implicated as a source of *C.jejuni*.

Pathogenesis: As few as 500 organisms can cause enteritis. The organism is invasive but generally less so than Shigella. Campylobacter produces adenylate cyclase-activating toxins same as of *E.coli* LT and cholera. **Incubation period:** Ranges from 2 to11 days.

Clinical features: Patients present with abdominal pain and cramps, diarrhea, malaise, headache, and usually fever. Typically the diarrhoea is watery, but in severe cases bloody diarrhea may occur. Diarrhea may last 2-7 days and the organism may be shed in the patients stool for up to 2 months. Bacteremia is observed in a small minority of cases. The disease is usually self-limiting.

Laboratory diagnosis: The feces may be inoculated in enrichment medium or on selective media such as Campy BAP or Skirrow's medium. The plates are incubated in microaerophilic conditions at 42oC for 2-5 days.

CHAPTER-17

SNAKE POISON

Introduction

Snake bites are encountered worldwide. Of the 3000 species of snakes, about 10% to 15% are venomous. Of the 14 families of snakes, 5 contain venomous species. Taiwan has about 23 venomous snakes. Snakes are poikilotherms, which account for their distribution and activity, and mostly active around 25~35°C. They distributed throughout most of the earth's surface, including fresh and salt water with only few exceptions.

Epidemiology

Reported snakebites are about 6000 annually in American and 2000 events are venomous bites. There are 300 to 600 reported snakebites in Taiwan annually, causing death of 20~30 people. 97% of snakebites are on the extremities. Males are bitten more frequently than females. 85% snakebites are predominate hematoxin.

Classification

There are about 23 types of venomous snakes in Taiwan, and 9 of them are sea snakes. Russell's vipera and Agkistrodon acutus are found over the southern and middle part of Taiwan. Others distrubted throughout the island.

Among five venomous families are

1. Colubridae
2. Hydrophidae / sea snakes: Hydrophis cyanocinctus Daudin Laticauda colubrina
3. Elapidae: Bungarus multicinctus (kraits), Naja Naja atra
4. Viperidae / true vipers: Vipera russelli formosensis (Russell's viper), Trimeresurus stejnegeri Schmidt, Trimeresurus mucrosquamatus (Turtle-Designed Snake)

5. Crotalidae / pit viper: Agkistrodon acutus / Deinagkistrodon acutus. Snake venom are classified as either neurotoxic or hematotoxic. Some of the snakes have compound toxin. Even this classification is felt inadequate currently, it is useful in clinical management.

Six of the most usually encountered snakebites in Taiwan are classified as below by venom toxicity.

Hematoxin: Crotalinae, Viperidae. Hematoxin alter vessel permeability, and procagulant toxin may cause swelling, severe pain, ecchymosis and comsuption coagulopathy. Agkistrodon acutus, Trimeresurus stejnegeri stejnegeri, Trimeresurus mucrosquamtus.

Neurotoxin: Elapidae. Neurotoxin acts on neuromuscular junction and causes paralysis. Signs usually become within 2~4 hours or may be delayed after the bite, which include ptosis, partial ophthalmoplegia, dysarthria, loss of facial expression, loss of airway control and respiratory paralysis.

Naja naja atra

Bungarus multicinctus

Mixed:

Vipera russell Formosensis

Sea snake

Identifications: Venomous vs. nonvenomous

Differentiation of between pit vipers and harmless snakes have several priciples. In the assessment of a reported bite from a venomous snake, one must distinguish the bite from that of a nonvenomous snake or another animal and from puncture wounds caused by inanimate objects(Figure).In the absence of positive identification, objective signs and symptoms of envenomation become the primary focus of diagnosis.

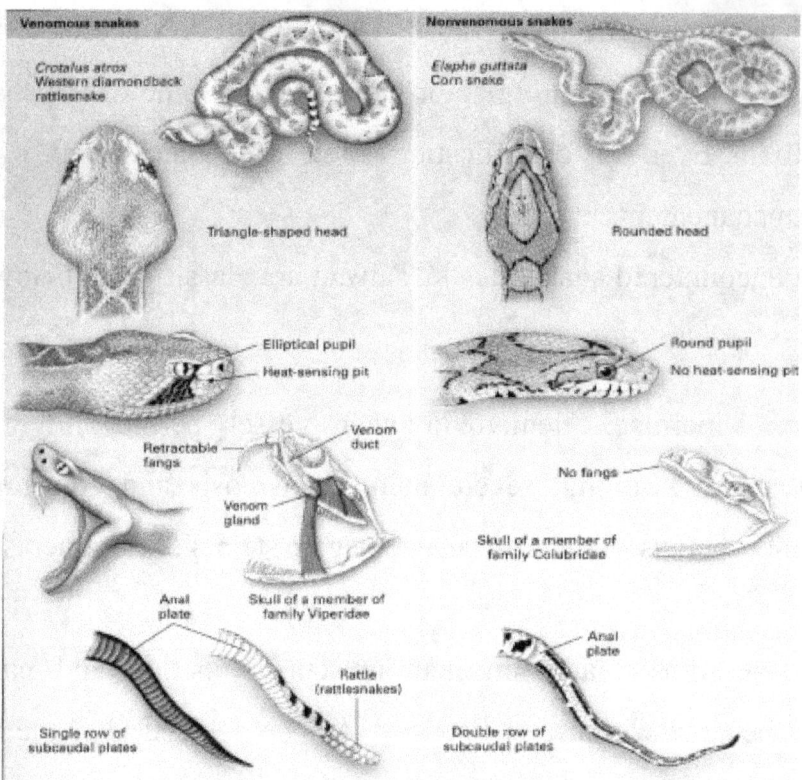

Figure 5: Differentiation between venomous and non-venomous snakes

Clinical Features

Observed snakebites is a straightforward diagnosis. Sometimes, highly suspecion of snakebites is required for vague bite mark or symptoms. Although suspected snakebite was common, severe envenoming occurred in less than 5% of cases. The signs and symptoms of a venomous snakebite vary considerably and depend on comorbidity, size, age of the victim; the age, health, size of the snake; relative toxicity of the venom; condition of the fangs; whether the snake has recently fed or its injured.

Local envenomation, if left untreated, can cause serious systemic problems as the toxic products are absorbed. The victim's autopharmacologic response to the Envenomation must also be taken into account. A wave of effects can occur over several days, ranging from minimal pain to multisystem failure and death.

From 30% to 50% of venomous snakebites result in little or no envenomation, less than 5% are severe envenomation.

Local effect. The fang mark from neurotoxin predominate snakebites, eg. Kraits or Naja naja atra usually are not remarkable with little pain or swelling and difficult to identify. Petechiae, edema, swelling and ecchymosis are remarkable from hematoxin predominate snakebites (i.e. pit viper). Severe localized pain and hemorrhagic bullae are also prominent. Oozing from fang marks usually is caused by Russell's viper. Necrosis of the skin and subcutaneous tissue is noted later. Numbness is the hallmark of neurotoxin snakebites. Duration from snakebites and accompanied soft tissue presentations are also useful in identifying snakebites. Progressive swelling of extremities may cause compartment syndrome and fasciotomy is usually required. Rhabdomyolysis and necrosis is inevitable.

Systemic effect. The most common reaction to snakebite is terror, which may cause nausea, vomiting, diarrhea, syncope, tachycardia, and cold, clammy skin. Many people believe that any bite from a venomous snake will result in envenomation; in fact, 25 percent of all pit-viper bites are "dry" and do not result in envenomation. Autonomic reactions related to terror must be differentiated from systemic manifestations of envenomation.

Some of the neurotoxin snakebites rapidly progressed to neuromuscular symptoms with dysphagia, dysarthria, general weakness, cranial nerves palsy, and even respiratory failure. Others may have prodromal phase of dizziness, nausea, vomiting, perioral numbness and tingling, metallic taste in the mouth, muscle fasciculation, ptopsis and then accompanied by acute respiratory failure, especially from *Naja naja atra*. The common cause of death is respiratory failure. Some of the systemic neuromuscular weakness may be delayed in onset, thus, close observation for at least 8 hours is required. Baseline and serial pulmonary function test are important.

Increased capillary membrane permeability and disruption of coagulation mechanism with bleeding are common in hematoxin predominating snakebites. Prolonged aPTT and PT, thrombocytopenia will not be reversed without antivenom therapy. Shock and massive pulmonary edema may be developed. Heart and kidney damage occurs secondarily to this mechanism. Some specific toxin may act directly on specific organ, such as heart, causeing heart failure and even arrythmias. Allergic reactions secondary to histamine and bradykinin release may aggravate systemic response.

Diagnosis

Snakebites are clinically suspected by history of exposure, clinical symptoms and laboratory tests. Snakes recognition may be assisted by pictures or specimen provided at emergency department. Fang marks, local or systemic symptoms are helpful.

(1) History of exposure (Snakes type, fang marks, sites of accidence, time, initial management.)

(2) Physical examinations (local effects: fang marks, swelling, ecchymosis, pain; systemic symptoms)

(3) Laboratory data (CBC, coagulation profile, chemistry, urinalysis)

Treatment

Medical management for snake bites includes first aids, emergency care, and antivenom therapy and monitoring possible complication.

First aids.

1. Try to recognize the snake by appearance, color and characteristics
2. Victim should be moved beyond striking distance to prevent second strike.
3. Placed at rest, kept calm and warm, as soon as possible. Keep physical activity minimal. Movement(i.e. walking) willfacilitatevenom absorption.
4. Rings, watches, and constrictive clothing should be removed on involved extremities, and stimulants (eg. alcohol) should be avoided. Immobilize extremitites in a functional position below the level of the heart by compressive dressing and splint. Constriction bands (elastic bandage or penrose drain, rope, or piece of clothing) may be of some use, especially when immediate medical care is not available. For hematoxin predominant snakebite (i.e. marked swelling and ecchymosis) wrapped circumferentially above the bite, applied with enough tension to restrict superficial venous and lymphatic flow while maintaining distal pulses and capillary filling. The band should be snug but loose enough comfortably slide a finger underneath. For neurotoxin

predominate snakebite, the bitten extremities should wrap in a snug elastic bandage, elastic bandage are applied initially over the bite site and then extending to cover the entire limb.

5. Previously recommended first-aid measures such as tourniquets, incision and suction, cryotherapy (ice water immersion), and electric-shock therapy are strongly discouraged. If a tourniquet or constriction band has been placed as first aid, it should be left in place until the victim is evaluated in the hospital and, if appropriate, until infusion of antivenom is initiated.

6. Patients are transported immediately to the nearest medical facility regardless of whether overt signs of envenomation are quickly apparent. Signs and symptoms of snakebites may be delayed.

7. Paramedical personnel should focus on support of the airway and breathing, administration of oxygen, establishment of intravenous access in an unaffected extremity, and transportation of the victim to the nearest medicalfacility

Emergency management. Any suspected snakebites should prompt the initiation of first aid, investigation and observation. Early consultation is recommended due to complex presentations. The first priority is mantaining of vital signs and advanced life support. If snakebites has been confirmed or highly suspected, the next step is to identify dry bite to envenomation. Envenomation grading is helpful to determine the need for antivenom.Prompt antivenom therapy, aggressive supportive resuscitaion and treatment of complications greatly reduce mortality of snakebites. The time elapsed since the bite allows assessment of the temporal effects of the bite to determine if the process is confined locally or if systemic signs have developed. Sign of envenomatinon should be aggressive sought, by clinical and laboratory evidence of venom effect. Assess the timing of events and onset of symptoms. Inquire about the time the bite occurred and details about the onset of pain. Early and intense pain implies significant envenomation

Absent: removed elastic bandage, observed for at least 12h. Delayed Envenomation after benign presentation has been documented, but most envenomation patients will develop clinical or laboratory evidence evidence of envenomation within 2h of removing bandage. Coagulation studies should be repeated 2h after bandageremoval and at interval therafter, depending on patient's condition.

Present: The bandage should be left in place until signs are absent or antivenom has been applied. If the patient's condition deteriorates immediately after bandage removal, the bandages may be reapplied while antivenom is given. Assess clinical severity for possible antivenom. The wound should be cleansed with soap and water. Immunized with tetanus. Wound culture and antibioitic therapy should be initiated only if signs of infection are present. Prophylactic antibiotics are not indicated in the routine treatment of patients with snakebites from neither non-venomous nor venomous snakes if no necrosis is present. Use of antibiotics prophylacticaly will have little impact on further infection but may give rise to side effects. It is not cost effective and may select out more resistant organisms. The event rate for infection after snake bite from venomous snakes is low.

Lab:
- CBC with manual differential and peripheral blood smear
- Prothrombin time and activated partial thromboplastin time
- Fibrinogen and split products
- Type and cross
- Blood chemistries, including electrolytes,
- BUN, creatinine, CK
- Urinalysis for myoglobinuria
- Arterial blood gas determinations for patients with systemic symptoms
- Baseline and serial pulmonary function parameter if suspect neurotoxin.

Antivenom. Antivenom is the mainstay of therapy for poisonous snakebites. Envenomation grading is helpful to determine the need for antivenom. Progression of signs and symptoms also indicated for antivenom therapy even after several days of snakebites. Antivenom is most effective if infusion within 4h, and less effective if more than 8 hours. Nevertheless, in severe envenomation, antivenom should be considered even after 3~4 days. Never use intramuscular or digital injection. Observation for progression of edema and systemic signs should be continued during and after antivenom infusion. Limb circumference should be measured at several sites above and below the bite. Repeat above eabs erery 4 hours.

There are four types of antivenom available.

1. Polyvalent hematropic antivenom;
2. Polyvalent Neurotrpic antivenom;
3. Monovalent antivenom for *A. acutus*;
4. Monovalent antivenom for Russell viper.

Progression of signs and symptoms: worsening of local injury: pain, ecchymosis, swelling. laboratory abnormalities: worsening platelet counts, prolonged coagulation times, decreased fibrinogen. systemicmanifestations: unstablevital signs or abnormal mental status. According to Dr. Dart's data, enovenomation can be validated by severity scoring. The scoring system depends upon minimal-moderate-severe score, grade I-IV score, and snake severity score. The total score interpretation is 1.3-2.8 for no risk, 2.1-2.3 for minimal risk, 3.2-3.5 for moderate risk, and 8. 5-9.5 for severe risk.

Complications or side effects from antivenom therapy.

Acute and delayed allergic reactions are fewer with new antivenom than conventional equine-derived antivenom. Acute reactions occurred in 19% of patients with isolated urticaria, cough, hives, dyspnea and wheezing. Infusion should be stopped immediatelyfor all allergic reactions. Mild cases may resolve spontaneously without treatment or responsed to antihistamine and steroid. In mod erate cases, continued antivenom infusion was given if the reaction response to antihistamine and steroid or the envenomation is very severe. Epinephrine infusion should be readily available depending on the severity of the reaction. Delayed serum sickness will be developed in 23% of patients usually begin from 2 to 10 days after antivenom administration and last for a week or more.. These include pruritis, rash, arthralgia, anorexia, and hives. But these are usually resolvedwithantihistamine(eg. Diphenhydramine, hydroxizine, cimetidine) or steroid (methylprednisolone) according to the severity of allergic reactions. Serum sickness is the only indication for the use of steroids with snakebites.

Complications of Envenomation

Coagulopathy. Antivenom is the best treatment for coagulopathy, but if active bleeding occurs, blood component replacement may be necessary.

Compartment syndrome. Severe envenomations may be associated with increased compartment pressure. The local reaction to envenomation, manifested as marked swelling, tenderness, tenseness, hypesthesia, and pain, may mimic a true compartment syndrome. In cases of suspected compartment syndrome, clinical diagnosis requires objective evidence of elevation in compartment pressure to more than 30 mmHg. If compartment pressure is elevated, we recommend elevation of the bitten body part in conjunction with the administration of mannitol 1~2g/kg IV over 30min and an additional four to six vials of FabAV over the course of one hour. Compartment syndrome in patients with envenomation is thought to be caused by myonecrosis related to the action of the venom components rather than the elevated compartment pressure that causes vascular insufficiency. Thus, the most effective treatment is to neutralize the venom, which may reduce the compartment pressure. If these measures fail to reduce compartment pressure within four hours and the patient has circulatory compromise, fasciotomy may be required to lower the compartment pressure. There is some debate regarding the use of fasciotomy, and no firm evidence support its usage. It does not prevent the progression of envenomation, treat coagulopathy, or obviate the need for additional antivenom. Fasciotomy may substantially lengthen the course of treatment and may be associated with nerve damage, disfiguring scars, contractures, and loss of limb function.

Rhabdomyolysis. May be a complication of compartement syndrome or caused by venom myolysins. Results in muscle pain, weakness, myoglobinuria, renal failure, and hyperkalemia. Alkalinization of urine and closely monitor urine output (ie hydration, mannitol or diuretics) is pertinent to prevent acute renal failure.

Summary

Suspected snakebites should prompt immediate first aid in field and early transportation to nearest medical facilities. Emergency physicians base on history, clinical manifestation and labo ratory data to confirm suspected snakebites. For dry bites or no envenomation, the patient should be observed for at least 6~12h before discharged. Extended period of observation to 12~24h is required for a coral snake. The EP should determine the severity of Envenomation and predominate venom activity before decide what type of antivenin to administer, how much, and over what period. Allergic reaction is not rare in antivenin therapy.

CHAPTER-18

HEAVY METAL TOXICITY

There are 72 metals in the periodic table. Those which are toxic to man are:
- Mercury (Hg)
- Lead (Pb)
- Cadmium (Cd)
- Arsenic (As)
- Iron (Fe)
- Aluminium (Al)
- Thallium (Th)
- Plutonium

Heavy metals are found everywhere: In our food, the air we breathe and in our water

Sources of heavy metals

1. Industrial and occupational
    a. Hg in paper production
    b. Hg and Pb in paint and chemical industry
    c. Hg is an old industrial poison - the term "Mad as a hatter" probably reflects what happened to people in the felt industry in France who used mercuric nitrate to give their felt hats a nice shine.
2. Agricultural
    a. Hg in fungicides
3. Household
    a. Hg and Cd in fish (particularly bottom dwelling fish since the heavy metals sink to the bottom) and shell fish
    b. Pb in paints and pottery glazes

4. Medical

    a. Hg in amalgam fillings (controversial)

5. Environmental

    a. Burning of leaded petrol (contains tetraethyl lead, an organometal)

Chemical forms of heavy metals

1. Elemental e.g. Hg, Pb
2. As inorganic compounds (salts) e.g. $MgCl_2$, $PbSO_4$
3. As organometallic compounds e.g. $(C_2H_5)Pb_4$ - Tetraethyl lead (A metal which is bound directly to the carbon of an organic compound)

Absorption of heavy metals

Absorption depends on the chemical form it is in:

1. Elemental: Normally poorly absorbed regardless of route (although Hg vapour, obtained by breaking a mercury thermometer, is effectively absorbed by inhalation)

2. Inorganic compounds: Depends on the solubility of the compound, Soluble salts are well absorbed, insoluble salts are poorly absorbed

3. Organometallic compounds: The increased lipophilicity will improve the passage across membranes. Hence these compounds can be effectively absorbed through the skin and can pass through the blood brain barrier. Once in the body, organometallic compounds can be metabolised to give the free metal ion (i.e. it can behave as an inorganic salt) or can remain stable and exert its effects as an intact organometallic compound.

Distribution of heavy metals

- Heavy metals will distribute (and have effects on) a variety of target organs. e.g. skin, gut, liver, kidney, haemopoietic system. central and peripheral nervous system
- May enter cells via ion transport mechanisms
- Organometallic compounds may cross directly through the BBB and accumulate in the CNS

Elimination of heavy metals

Heavy metal toxicity often persists in the body because: Unlike organic compounds, heavy metals cannot be metabolised. The only method of elimination from the body is via excretion. Heavy metals bind effectively to macromolecules and hence their excretion is slow (half life = days, weeks)

Excretion
- Takes place predominantly from the kidney
- Also from:
  - Sloughing of skin
  - Hair and nail growth
  - Breath (Hg)

Types of heavy metal intoxication

1. Acute: Result of exposure to very high levels of metal (industrial accidents)
2. Chronic: Low level exposure which accumulates over time

Symptoms of intoxication

Depends on the compound and the chemical form it is in (consider lead as an example)

1. Elemental and inorganic salt form.
   - Acute: Local irritation of the gut
   - Chronic symptoms may develop if sufficient lead is absorbed
   - Chronic:
     - GIT - Anorexia, dyspepsia, constipation
     - Haemopoetic system - Anaemia, basophilic stipling
     - Neurological - Peripheral neuropathy, wrist and foot drop, lead encephalopathy
     - Renal - Albuminuria
     - Oral - Ulcerative stomatitis, lead line (deposition of Pb in gums)

o   Reproductive - Spontaneous abortion, sterility
2. Organomettalic form
- Major effects are in the CNS
- Neurologic symptoms
- Psychiatric symptoms
- No lead lines, no haemopoeitc effects

Mechanism of heavy metal toxicity

Heavy metals are electrophilic and react with nucleophilic groups on proteins in the body. These nucleophilic groups are: -SH, -NH$_2$, -CONH. -PO$_4^-$. As a result of binding to these groups on proteins. the heavy metal may compete with endogenous metals, e.g. Ca^{2+} or Zn^{2+} which also rely on binding to these proteins via the same groups.

Heavy metal binding may lead to disruption of enzyme function. e.g. Pb interferes with haem biosynthesis by inhibition of 8 aminolevulinic acid dehydratase (ALAD). This results in a reduction in plasma ALAD activity and an increase in plasma aminolevulinic acid.

Treatment of heavy metal toxicity

1. Removal of the source of the toxicity
2. If ingested, alleviation of the acute presence
   a. Emesis
   b. Lavage
   c. Charcoal
3. Support
4. Volume repletion
5. Reducing body stores - via chelating agents

Chelating agents

Contain 2 or more electron donating groups (eg. N or O) which form coordinate bonds to the metal ions. Our body contains some natural chelating agents: Vitamin B12 (chelates Co), Haemoglobin (chelates Fe), Chlorophyll in plants (chelates Mg), cP450 (chelates Fe, Cu). By binding to the metal, the chelating agent enhances excretion of the heavy metal. Thus, they must be:

1. Stable when administered
2. Form stable complexes with the metal that needs to be removed
3. Have greater affinity to the metal than the metal has to biological groups on proteins
4. Be readily excreted (e.g. by kidneys)

Examples of chelating agents: Dimercaprol, Ethylene diamine tetracetic acid (EDTA), Penicillamine

Dimercaprol

Good for acute intoxication with Hg, As, Au. Used in combination with calcium EDTA in treatment of lead intoxication. Will reactivate enzyme systems. The metal mercaptide complex which forms, although stable, can redissociate to release the bound metal, hence it is not very effective

Pharmacokinetics:

Poorly absorbed after oral administration. Given as a deep intramuscular injection in oil. Maximum serum levels after 50 to 60 min. Hence, it takes a long time for the agent to dissociate away from the muscle and into the blood stream. It can distribute to all tissues and cross the BBB. Excreted in bile (hence it can be used in patients with renal impairment).

Dosage:

For acute lead encephalopathy, dimercaprol is given in combination with calcium EDTA. The complex is unstable at low pH, so it is necessary to keep the urine alkaline (e.g. with $NaHCO_3$)

Side effects:

- CNS - vomitting, tremor, seizures
- Cardiac - hypertension, tachycardia
- Burning sensation around lips, mouth, throat, penis

Calcium EDTA

Will chelate any metal which has a greater affinity than Ca^{2+}, i.e. Ca^{2+} is displaced from EDTA by the heavy metal which then forms a stable complex. The complex is excreted in the urine. The presence of Ca^{2+} allows rapid IV administration without depletion of endogenous Ca^{2+} were not present nitially, the injected EDTA will chelate endogenous Ca^{2+} first. It is the treatment of choice for lead poisoning and also useful for: Cr, Mn, Ni, Pu, Th, Ur. Not effective for: Hg, Au, As. It should not be given orally. This is because lead in the gut forms soluble complexes with EDTA which is readily absorbed.

Pharmacokinetics:

Half life: 20-60 min if given IV, 1.5 hours if given IM. Mostly excreted in the urine unchanged by glomerular filtration. EDTA does not enter cells. Initially, there is a peak rise in Pb chelated with EDTA in the urine. This is Pb which is weakly bound in the body. When this Pb has been mopped up, there is a decline in Pb-EDTA in the urine. Intracellular Pb then moves out of the cells (due to the reduction in extracellular Pb, i.e. Pb is moving down its concentration gradient). This intracellular Pb, after it comes out of the cells, is chelated with EDTA, to cause a second peak rise in Pb-EDTA complex in the urine.

Dosage:

Adequate urine flow must be established prior to use. For acute lead intoxication, administer dimercaprol concurrently.

Side effects:

Arrythmias, mild histamine like effects, hypercalcaemia due to the Ca^{2+} which is initially bound to EDTA when administered being displaced.

Penicillamine

Effective chelator of Cu, Hg, Fe. Zn, Pb and Au. Used in the treatment of hepatocellular degeneration in Wilson's Disease (excessive Cu). Rheumatoid arthritis (it is a SAARD) – Less likely to be used now (updated, 2000) due to more effective use of Methotrexate as a disease modifying agent.

Pharmacokinetics:

Can be administered orally. Peak serum levels 1 hour after administration. Rapidly excreted in urine

Dosage:

30mg/kg/day or 250mg 4 times daily

Side effects:

- Hypersensitivity
- Nausea, vomiting
- Leukopenia, neutropenia
- Inhibits pyridoxal dependent enzyme systems, hence dietary pyridoxal supplements are given

CHAPTER-19

ORGANOPHOSPHOROUS POISONING

Acute organophosphorous poisoning (OPP) occurs following dermal, respiratory or oral exposure. Organophosphorous compounds (OPCs) can be classified into low volatile compounds eg. chlopyriphos, dimethoate, dichlorvos, methyl parathion etc. used for agricultural purposes as pesticides or highly volatile nerve gases eg. sarin, tabun etc., mainly used in chemical warfare. Most cases occur in developing countries and are generally following suicidal ingestion. World Health Organization (WHO) has estimated that nearly 200,000 people worldwide die from pesticide poisoning mainly in developing countries following intentional poisoning. This is because of their wide and easy availability and occupational exposure because of inadequate or inappropriate protective equipment. Military and terrorist attacks with nerve gases always remain possible eg. Iran-Iraq war, Tokyo underground attack.

In India, it is the commonest poisoning. OPCs inhibit acetylcholinesterase at neuromuscular junction, in autonomic and central nervous system resulting in accumulation of acetylcholine (ACh) and over stimulation of ACh receptors resulting in acute cholinergic crisis which is characterized by bradycardia, bronchorrhoea, miosis, sweating, salivation, lacrimation, defecation, urination and hypotension. In addition, there occurs muscle weakness and fasciculations. The CNS involvement results in alteration in sensorium and seizures. Following resolution of cholinergic crisis, some patients may develop intermediate syndrome i.e. cranial nerve palsies, proximal muscle weakness, respiratory muscle weakness. Some may develop peripheral neuropathy (OPIPN) at a later stage.

The diagnosis can be made from history of ingestion or exposure e.g. following spray, clinical features and plasma cholinesterase (PChE) and red cell acetyl cholinesterase (Red Cell AChE) inhibition. However, between inhibition of these enzymes and severity of poisoning there is no correlation. The management of these patients involves washing of skin and induction of vomiting or gastric lavage to remove OPCs from skin and stomach, administration of activated charcoal, atropine, glycopyrrolate, oximes and some newer compounds in addition to ventilatory support which they may require.

Washing of skin and removal of contaminated clothes No randomized controlled trials (RCTs) are available in literature. However, it seems to be the most obvious way of reducing further dermal and mucosal absorption. However, care should be taken by health workers to protect themselves by using gloves, aprons, eye protection etc. as they run the risk of getting poisoned. Moreover, this should not be priority if patient requires resuscitation first. Induction of vomiting with ipecac No RCTs are available in literature. However, complications have been reported following its use. These are aspiration, diarrhoea, ileus etc. One systematic review suggests that the use of ipecac in any poisoning does not improve the outcome. Moreover, administration of it is likely to result in delay in administration of activated charcoal.

Gastric lavage

The complications include aspiration, laryngeal spasm, oesophageal perforation, hypoxia. These are especially common when it is being performed in a struggling, non-consenting patient. Although anecdotal reports suggest that OPCs may remain in gut for prolonged duration and it may help in their removal, there is no obvious evidence at present that it helps in outcome. In India, as suicide is still an offence and gastric lavage is being done routinely, for medico-legal reasons to collect gastric sample and for therapeutic reasons, it will be better to carry out RCTs to see whether it benefits the patients.

Activated charcoal

The complications include aspiration pneumonia, vomiting, diarrhoea, constipation, ileus and reduction in absorption of oral medications. There are few trials which suggest that incidence of complications is low with multiple dose regimen. One non-systemic review of single dose of activated charcoal in all forms of poisoning has found that it does not improve the outcome. There is a need to carry out RCTs to find benefit of single or multiple dose regimen in patients with OPP.

Atropine

Atropine remains the main stay of treatment. Although it has not been compared with placebo, several case series have found that it reverses the early muscarinic effects of OPP. Atropine competes with excess ACh at muscarinic receptors. The first doses are generally given as boluses followed by infusion if dose requirement is large. The rate of infusion should be kept to maintain pupils at midpoint, heart rate greater than 100 beats per min, normal bowel sounds and clear lungs.

Glycopyrrolate

Glycopyrronium bromide has been used in place of atropine. In a small RCT comparing it with atropine (total 39 patients), it was found that there was no significant difference in the outcome in two groups except that fewer respiratory infections were observed in patients who were given glycopyrrolate and were ventilated. The mortality rate and duration of ventilation did not differ. However, a major limiting factor is the cost. Glycopyrrolate is about 10 times more costly and it may be worthwhile to carry out more RCTs to know whether it has any benefit over atropine.

Oximes

Oximes reactivate the acetylcholinesterase inhibited by OPCs, reactivation is limited by ageing and by high concentration of pesticide. With diethyl compounds ageing takes longer than with dimethyl compounds. Complications of oximes include hypertension, cardiac dysrhythmias, headache, blurred vision, dizziness etc. Obidoxime can lead to hepatic failure. Two RCTs are available from Vellore (India), suggesting that oximes do not benefitand with 12 gm over 3 days increase the risk of death, intermediate syndrome and requirement of ventilation. However, the studies have been criticized for randomization bias and inadequate dose. It is suggested that 2-PAM infusion should be given till patient recovers. WHO currently recommends

30 mg/kg bolus followed by 8 mg/kg/hr as IV infusion. RCTs MJAFI, Vol. 60, No. 1, 2004 involving large number of patients are required to prove its benefit.

Newer compounds

Organophosphorus hydrolases such as mammalian paraoxonase can hydrolyze the OPCs thus reducing their concentration rapidly. However, no human studies are available.

Sodium bicarbonate

Animal studies suggest that increasing pH with sodium bicarbonate may reduce mortality rate and this effect is independent of acidosis. At present, a few uncontrolled studies are available to show its benefit but no RCTs are available.

Clonidine

It inhibits the release of ACh from cholinergic neurons and has adrenergic agonist effects. In animal studies, pre-treatment with it improves survival. However, no human studies are available. N-methyl-D-aspartate receptor antagonists (NMDA receptor antagonists) : Primate studies have found that pre-treating with NMDA receptor antagonists such as gacyclidine improves recovery. However, no human studies are available.

Benzodiazepines

Diazepam is the standard treatment for organophosphorous induced seizures. No RCTs are available but several studies support that diazepam controls seizures.

Conclusion

Washing the patients and removing the contaminated clothing, administering atropine to control muscarinic manifestations and diazepam to control seizures are undoubtedly of use in management of acute organophosphorous poisoning. Induction of vomiting with ipecac may prove more harmful. Gastric lavage, at present, seems to lead to more harm than benefit especially in a struggling, non consenting patient. Oximes need to be studied in larger RCTs to find the benefit, using the recommended doses. There is no evidence at present that organophosphate hydrolases, sodium bicarbonate, clonidine, NMDA receptor antagonists help in outcome.

All the best to my dear students

www.ingramcontent.com/pod-product-compliance
Lightning Source LLC
Chambersburg PA
CBHW080921170526
45158CB00008B/2189